73rd Annual Meeting of the Northeastern Weed Science Society 2019

Proceedings – Volume 73

Hunt Valley, Maryland, USA
8-10 January 2019

ISBN: 978-1-5108-8581-3

Printed from e-media with permission by:

Curran Associates, Inc.
57 Morehouse Lane
Red Hook, NY 12571

Some format issues inherent in the e-media version may also appear in this print version.

Printed by Curran Associates, Inc. (2019)

For permission requests, please contact Northeastern Weed Science Society (NEWSS)
at the address below.

Northeastern Weed Science Society (NEWSS)
P.O. Box 25
Woodstown, NJ 08098
USA

Phone: 315-209-7580

northeasternweedscience@hotmail.com

Additional copies of this publication are available from:

Curran Associates, Inc.
57 Morehouse Lane
Red Hook, NY 12571 USA
Phone: 845-758-0400
Fax: 845-758-2633
Email: curran@proceedings.com
Web: www.proceedings.com

PROCEEDINGS - VOLUME 73

73nd Annual Meeting of the

Northeastern Weed Science Society

January 8-10, 2019 – Delta Hotels Baltimore Hunt Valley, Hunt Valley, MD

Site of 2020 Joint Meeting*

January 5-8, 2020

Hilton Philadelphia at Penn's Landing
Philadelphia, PA

Participating Societies:
Northeastern Weed Science Society
American Society of Agronomy – Northeastern Regional Branch
American Society of Horticultural Science – Northeast Division

To order a copy of the Proceedings contact:
CURRAN ASSOCIATES, INC
57 Morehouse Lane, Red Hook, NY 12571
Ph: 1 845 758 0400 x112

§ Cover: The winning image in the 2019 NEWSS Photo Contest, submitted by Thierry Besançon, Rutgers University.

Proceedings
of the
Seventy Third Annual Meeting
of the
Northeastern Weed Science Society

Michael L Flessner, Editor

NORTHEASTERN WEED SCIENCE SOCIETY

2019 SUSTAINING MEMBERS

Platinum

BASF — We create chemistry

syngenta

Gold

FMC

VALENT®

Bayer CropScience

Silver

UPL

Bronze

ACDS Research, Inc.

TeeJet Technologies

LABServices

For-Shore Weed Control

Table of Contents

NEWSS EXECUTIVE COMMITTEE – OFFICERS

President	**Dan Kunkel** Rutgers University kunkel@aesop.rutgers.edu
President-elect	**Art Gover** The Pennsylvania State University aeg2@psu.edu
Vice President	**Caren Schmidt** BASF caren.schmidt@basf.com
Treasurer	**Kurt Vollmer** University of Delaware kvollmer@udel.edu
Past President	**Carroll Moseley** Syngenta Carroll.moseley@syngenta.com

NEWSS EXECUTIVE COMMITTEE – MEMBERS

Membership Chair	**Thierry Besançon** Rutgers University thierry.besancon@rutgers.edu
Editor	**Michael Flessner** Virginia Tech flessner@vt.edu
Public Relations	**Katelyn Venner** BASF katelyn.venner@basf.com
Research and Education Coordinator	**Matt Cutulle** Clemson University mcutull@clemson.edu
Sustaining Membership	**Wesley Everman** North Carolina State University wes_everman@ncsu.edu
CAST Representative	**Hilary Sandler** University of Massachuesetts Cranberry Station hsandler@umass.edu
Graduate Student Representative	**Theresa Piskackova** North Carolina State University tareinha@ncsu.edu
WSSA Representative	**Rakesh Chandran** West Virginia University rschandran@mail.wvu.edu
Science Policy Director	**Lee Van Wychen** Weed Science Society of America lee.vanwychen@wssa.net

2019 NEWSS MEETING PROGRAM COMMITTEES

Program Committee	Chair:	Dan Kunkel, Rutgers University Art Gover, The Pennsylvania State University
Agronomic Crops	Chair:	Larissa Smith, Syngenta
Vegetation Management and Restoration	Chair: Chair elect:	Lane Heimer, Maryland Dept. of Agriculture Art Gover, The Pennsylvania State University
Turfgrass and Ornamentals	Chair: Chair elect:	Matt Elmore, Rutgers University Jatinder Aulakh, The Connecticut Agricultural Experiment Station
Fruit, Vegetable, and Tree Crops	Chair:	Ramón Arancibia, Virginia Tech
Research Posters	Chair:	Kate Venner, BASF
Student Oral Papers	Chair:	Shawn Askew, Virginia Tech

2019 NEWSS COMMITTEES

Student Paper Contest	Chair:	Carroll Moseley, Syngenta
Student Poster Contest	Chair:	Matt Elmore, Rutgers University
Photo Judging Contest	Chair:	Hilary Sandler, University of Massachusetts
Site Selection Committee	Chair:	Art Gover, The Pennsylvania State University
Awards Committee	Chair:	Carroll Moseley, Syngenta
Nomination Committee	Chair:	Randy Prostak, University of Massachusetts
Resolutions Committee	Chair:	Jerry Baron, Rutgers University
Collegiate Weed Contest	Chair:	Art Gover, The Pennsylvania State University
Audit Committee	Chair:	Kurt Vollmer, University of Delaware
Archives Committee	Chair:	Dan Kunkel, Rutgers University
Weed Science Field Days	Chair:	Quintin Johnson, University of Delaware
Past Presidents Committee	Chair:	Carroll Moseley, Syngenta
Herbicide Resistance	Chair:	Mark VanGessel, University of Delaware
Endowment Committee	Chair:	Art Gover, The Pennsylvania State University

2019 AWARDS, NEWSS ANNUAL MEETING

NEWSS Fellow Award – Antonio (Toni) DiTommaso

Award of Merit - Matthew J. Mahoney

Outstanding Researcher Award – Michael Flessner

Outstanding Educator Award - Hilary A. Sandler

Robert D. Sweet Outstanding Ph.D. Student Award – Shawn C. Beam

Robert D. Sweet Outstanding M.S. Student Award – Kara B. Pittman

M. Garry Schnappinger Service Recognition Award – Erin M. Hitchner

2019 NEWSS MEETING CONTEST WINNERS

STUDENT PAPER

1st place paper – Eric Jones, North Carolina State University, "Determining Sorgoleone Efficacy on Select Wheat (*Triticum aestivum* ssp.) Cultivars and Weed Species." (9)

2nd place paper – Katie Diehl, Rutgers University Annual Bluegrass Weevil (*Listronotus maculicollis*) and Paclobutrazol for Annual Bluegrass (*Poa annua*) Control in Fairways." (5)

Honorable mention – D.G. Lewis, Clemson University, "Interaction of Organic Herbicides and Cover Crops on Weed Suppression and Insect Dynamics in No Till Systems." (2)

STUDENT POSTER

1st place: John Brewer, Virginia Tech, "Bulbous Bluegrass: A Weed of Ornamental Turf" (3)

2nd place: Theresa Reinhardt Piskackova, North Carolina State University, "Modeling Palmer Amaranth (*Amaranthus palmeri*) Emergence Using Chronological, Thermal, and Hydro-thermal Time." (6)

Honorable Mention: D.J. Contreras, North Carolina State University, "Critical Period of Weed Control for Ryegrass in Wheat" (5)

Honorable Mention: D.G. Lewis, Clemson University, "Evaluation of Weed Size and Water Carrier pH on Suppress Herbicide Efficacy." (1)

PHOTO CONTEST

1st place photo – Thierry Besançon, Rutgers University, Bloom of red clover

2nd place photo – Dwight Lingenfelter, Penn State, Thistle in the landscape

2018 NEWSS COLLEGIATE WEED SCIENCE CONTEST
Hosted by BASF, Pine Level, NC

(Photos courtesy of Katelyn Venner)

The 2018 NEWSS Annual Weed Contest was hosted at BASF Corporation in Pine Level, NC, on July 25, 2018. Caren Schmidt from BASF organized the event and with help from many others provided an amazing weed contest. Thirteen graduate teams, nine undergraduate teams and two individuals competed for a total of 88 students from seven universities across the region. The contest went smoothly despite threats of heavy rain and storms. Luckily, it only rained a little bit and the contest was a success! The contest concluded with the awards ceremony at Hinnant Family Vineyards, where students were given a taste of muscadine grape juice made at the vineyard. The contest was quite competitive, Virginia Tech Graduate Team 1 won the graduate student division and University of Guelph Undergraduate Team 1 won the undergraduate division. BASF put on a challenging contest, and it was a resounding success. They could not have done this without support from management, and each and every volunteer.

Farmer Problems

At the Vineyard

Hinnant Family Vineyards, Pine Level, NC

After a long day, everyone is cleaned up and enjoying good food and company.

A proud Caren Schmidt (BASF) hands out the awards to the highest performing teams and individuals.

Ryan Aldrich (BASF) walks the students through Weed ID.

Shawn Askew can't seem to get rid of two of his favorite students, Sandeep Rana and Kate Venner.

Awards

Undergraduate Team Awards

First Place Undergraduate

Guelph Undergraduate Team 2

(Christena Jenkins-Giffen, Bernard Wierenga, William Kell, Amy Rodenburg)

Second Place Undergraduate

Guelph Undergraduate Team 1

(Luke Hannam, Alan Abdulkader, Allison Bailey, Mackenzie Mahon)

Third Place Undergraduate

Guelph Undergraduate Team 3

(Bryce Ribey, Thomas Bowman, John Fluttert, Kyle Rutherford)

Awards... cont'd

Undergraduate Individual Awards

First Place Undergraduate

Allison Bailey (Guelph)

Second Place Undergraduate

Aleah Butler-Jones (Cornell)

Third Place Undergraduate

Jonathan Berlingeri (Cornell)

Awards… cont'd

Graduate Team Awards

First Place Graduate Team

Virginia Tech Graduate Team 1

(Shawn Beam, Jordan Craft, Lucas Rector, Wykle Greene)

Second Place Graduate Team

Guelph Graduate Team 1

(Lauren Benoit, Andrea Smith, Brendan Metzger, Todd Frey)

Third Place Graduate Team

Virginia Tech Graduate Team 3

(Harrison Ferebee, Hunter Blake, Eric Scruggs)

Awards… cont'd

Graduate Individual Awards

First Place Graduate Student

Shawn Beam (Virginia Tech)

Second Place Graduate Student

Lucas Rector (Virginia Tech)

Third Place Graduate Student

Jordan Craft (Virginia Tech)

EVALUATION OF WEED SIZE AND WATER CARRIER PH ON SUPPRESS HERBICIDE EFFICACY. D.G. Lewis*, M.A. Cutulle, T. Campbell, and D. Jeffris, Clemson University, Clemson, SC (1)

ABSTRACT

Weed control in organic vegetable production is limited; growers typically rely on cost and labor-intensive measures such as tilling and hand-weeding. Newly registered (2017) organic herbicide, Suppress EC, could reduce weeds as well as the amount of intensive measures needed for adequate weed control within these systems. Suppress efficacy was evaluated on varying weed type and height, as well as the effects of water pH and herbicide concentration in greenhouse and field settings.

UTILIZING UAVS AND SPECTRAL IMAGING TO DISTINGUISH SELECT WEED SPECIES IN A FIELD ENVIRONMENT. E.A. Jones*, J.T. Sanders, W.J. Everman, D.J. Contreras, and M.A. Granadino, North Carolina State University, Raleigh, NC (2)

ABSTRACT

For unmanned aerial vehicle (UAV) remote sensing to be employed as a viable and widespread tool for weed management, the accurate detection of distinct weed species must be possible through the use of analytical procedures on the resultant imagery. In 2017, two field studies were performed to identify any weed height thresholds on the accurate species detection and species by height classification of three common broadleaf weed species in Rocky Mount, North Carolina: Palmer amaranth (*Amaranthus palmeri*), common ragweed (*Ambrosia artemisiifolia*) and sicklepod (*Senna obtusifolia*). Pots of the three species at heights of 5, 10, 15, and 30 cm were randomly arranged in a grid and 5-band multispectral imagery was collected at 15 m above the ground. Image analysis was used to identify the spectral reflectance behavior of the weed species and height combinations and to evaluate the accuracy of species based supervised classifications. Supervised classification was able to discriminate between the three weed species with between 20-100% accuracy depending on height and species. Palmer amaranth classification accuracy was improved at larger weed heights and detection accuracy for the species was at least 68%. Increased height of sicklepod and common ragweed plants did not reliably confer improved accuracy but the species were correctly identified with at least 68% and 40% accuracy, respectively. Spectral separation was found to be dependent on the species, band, and height being observed. Reflectance of the species in bands 1, 2, 4, and 5 increased with height while band 3 reflectance declined as weed height increased. Finally, spectral separation was found to exist between the species, demonstrating the potential for using spectral reflectance measurements to discriminate between them.

BULBOUS BLUEGRASS: A WEED OF ORNAMENTAL TURF. J. Brewer* and S.D. Askew, Virginia Tech, Blacksburg, VA (3)

ABSTRACT

In the past few years, bulbous bluegrass (*Poa bulbosa*) has been regularly submitted to the Virginia Weed Clinic as a weed of lawns. It is a clump-forming perennial grass native to Eurasia and northern Africa, but can now be found in most of the United States. It typically emerges during winter and/or spring, but dies out as temperatures increase moving into the summer months. Previous research has shown it to compete with desirable vegetation in rangeland and in tall fescue lawns. Bulbous bluegrass reproduces asexually via bulblets on its inflorescences instead of seed. This reproduction strategy is unusual among grasses but similar to other monocots like wild garlic. These bulblets are viable without any period of dormancy and stay viable for at least 2 years. Due to its perennial nature and production of bulblets, bulbous bluegrass can be very difficult to remove from desired areas. The majority of research evaluating chemical control of bulbous bluegrass is centered in warm-season pastures and using ALS-inhibiting herbicides like sulfosulfuron, sulfometuron, and imazapic. Little is known about how the weed may be controlled in tall fescue. In cool-season turfgrass, research has been conducted to evaluate herbicides to control the related weed *Poa annua*. Some of the most effective options found for *Poa annua* include combinations of amicarbazone, ethofumesate, and mesotrione. At Virginia Tech, we established two greenhouse studies to evaluate different cool-season herbicides for bulbous bluegrass control. The plants were collected at a field site in January 2018 and maintained in greenhouse conditions for approximately one year before two studies were initiated on April 5 and April 18, 2018. The two studies were arranged in a randomized complete block design with three replications. All trials were sprayed with a CO_2-powered spray-chamber calibrated to apply 280 L ha^{-1}.

In spring 2018, we observed that amicarbazone at 105 g ai ha^{-1} and amicarbazone at 105 g ai ha^{-1} plus mesotrione at 280 g ai ha^{-1} controlled bulbous bluegrass greater than 80% during both studies and greater than all other treatments. Amicarbazone plus mesotrione completely controlled all plants while amicarbazone alone became inconsistent as plants were larger and more mature, but was still statistically similar to amicarbazone plus mesotrione. Normalized difference vegetation index, green cover assessed via digital image analysis, and final dry biomass all supported trends observed via visual ratings.

DICAMBA INJURY VARIES WITH SOYBEAN VARIETY AND MATURITY GROUP. M.A.

Granadino*, W.J. Everman, J.T. Sanders, D.J. Contreras, and E.A. Jones, North Carolina State University, Raleigh, NC (4)

ABSTRACT

In the last two decades, the use of dicamba has increased to control glyphosate-resistant weeds (Alves et al 2018). Dicamba-resistant crop varieties have the potential to become widely utilized as a tool to manage glyphosate-resistant weeds in North Carolina. Dicamba sprayed as low as 0.04 g ae h^{-1} has been reported to reduce yield by up to 10% (Weidenhamer et al 1989). Soybean has been found to be more susceptible to dicamba applications when exposed at flowering (R1 and R2) compared to vegetative stages (V1 to V7) (Kniss 2018). This research was conducted to evaluate injury of sub-lethal rates of dicamba in maturity group V and VI soybean cultivars at V4 and R2, and evaluate if visual injury is an accurate method of estimating yield loss. Two separate studies were conducted in 2018 in Kinston and Rocky Mount, NC. Study group was organized by soybean maturity group and planting timing (May and June). Plots consisted of two borders rows to minimize potential drift. Dicamba applications were made with a pressurized CO_2 backpack sprayer. Dicamba was sprayed at 0.44, 1.76 and 7.04 g ae a^{-1} at V4 and R2 at each respective planting date. Weekly visual injury ratings and height measurements as well as final yield were collected. Applications between June V4 and May R2 had no significant difference in injury. Dicamba rates had no significant effect on maturity groups and varieties' effect on yield, regardless of the planting date and application timing. There was no significant difference in yields or injury between maturity groups and varieties. Visual injury is not an accurate method of estimating yield loss.

CRITICAL PERIOD OF WEED CONTROL FOR RYEGRASS IN WHEAT. D.J. Contreras* and W.J. Everman, North Carolina State University, Raleigh, NC (5)

ABSTRACT

Winter wheat (*Triticum aestivum*) is often used in double cropping systems, as it can aid reducing soil erosion, weed suppression, and provides growers with an additional source of income. Italian ryegrass (*Lolium multiflorum*) is one of the most troublesome weeds for wheat production throughout Southern US. The inability to control this weed can result in reduced yields, reduced quality, or both. Cultural practices, such as the Critical Period of Weed Control (CPWC) can help reduce yield losses. The CPWC is a period in a crop´s growth cycle during which weeds must be controlled to prevent yield losses. The objective of this study was to determine winter wheat's critical period of weed control for Italian ryegrass that results on or less than 5% total yield loss. Plots were kept weedy or weed free through 2, 4, 6, 8, 10, 12, 14, 16, 18, 20 weeks after crop emergence. Two controls consisting of weedy and weed free all season plots were used for a treatment comparison based on a regression analysis where time of weed removal was related to crop yield. The CPWC to reduce yield losses to 5% or less was found to start around the 7th WAE (6.89, $p \leq 0.05$) and end at the 19th WAE (19.36, $p \leq 0.05$). Wheat most be maintained weed free during this period to prevent significant yield losses.

MODELING PALMER AMARANTH (*AMARANTHUS PALMERI*) EMERGENCE USING CHRONOLOGICAL, THERMAL, AND HYDRO-THERMAL TIME. T.A. Reinhardt Piskackova*, C. Reburg-Horton, R.J. Richardson, K.M. Jennings, and R.G. Leon, North Carolina State University, Raleigh, NC (6)

ABSTRACT

Predictive models for weed emergence can be important tools to meet the goals of Integrated Weed Management. While many studies have described germination requirements, no empirical seedling emergence model has been developed for Palmer amaranth. This study tracked Palmer amaranth emergence in Kinston, NC and Goldsboro, NC from March to September to develop such a model. Chronological, thermal, and hydrothermal time were used to model emergence patterns, and all these parameters were best fit to a Gompertz equation. The hydrothermal time model performed better than the thermal, which suggested that soil moisture conditions are an important variable to ensure the robustness of Palmer amaranth emergence predictions over a wider range of environmental conditions and soil types.

CONFIRMING GLYPHOSATE RESISTANCE IN HORSEWEED THROUGH SEED GERMINATION AND WHOLE PLANT ASSAY EVALUATION. C.S. Jennings* and A. Witcher, Tennessee State University, McMinnville, TN (7)

ABSTRACT

Glyphosate resistant horseweed (*Conyza canadensis*) is a significant issue in nursery crop production but has not been studied broadly. The objective of this project was to evaluate a leaf dip assay for resistance detection and determine the effectiveness of glyphosate rate on resistant and susceptible horseweed at three different growth stages. Two horseweed populations (resistant and susceptible) were identified, and individual plants were treated with glyphosate and glufosinate. Resistant plants survived glyphosate treatments and died following glufosinate treatments while all susceptible plants died. Seeds were collected from 10 plants within each population and seedlings (from each plant) were grown for the leaf dip and whole plant assays. For the leaf dip assay, seedlings were grown to the rosette stage and leaves harvested randomly from 16 seedlings/plant/population. Individual leaf samples (2.5 cm long from the tip) were harvested and placed into 7 ml tubes filled with glyphosate at 0, 300, 600, or 1,200 mg L^{-1}, then all samples were placed in a growth chamber at 26/21°C day/night cycle. After 72 hours, leaves were scanned and analyzed with WinFOLIA software to calculate percent healthy tissue. For the whole plant assay, plants were grown to three different growth stages (seedling, rosette, and bolting) and treated with glyphosate at 0 (control), 2.34 (low), 11.69 (medium), or 23.38 (high) L/hectare. After 21 days, plant shoots were harvested and weighed (fresh weight). The percentage of living plants was also determined at time of harvest.

In the leaf dip assay, the resistant population had 94, 89, 81, and 69% healthy tissue (control, low, medium and high rates, respectively). Control and low rates were similar, yet the medium and high rates were significantly lower than the control. The susceptible population had 97, 67, 43, and 40% healthy tissue (control, low, medium and high rates, respectively). All glyphosate treatments had significantly lower healthy tissue compared with the control. Percent healthy tissue decreased by 57% within the susceptible population from the control to the high treatment compared to only a 25% decrease within the resistant population. Percent healthy tissue in the resistant leaves was 69% at the high rate while susceptible leaves had only 67% healthy tissue at the low rate. Based on the results, the leaf dip assay can be used to distinguish between resistant and susceptible horseweed plants. In the whole plant assay, the resistant fresh weight was similar in the control, low and medium rates. In the susceptible population, fresh weight significantly decreased with each increase of glyphosate when compared with the control. The percentage of living plants was significantly different between resistant (greater percentage living) and susceptible (lower percentage living) populations for each treatment except for the control. Within the susceptible population, survival was similar for the control and low rate at each growth stage except seedling. In conclusion, glyphosate resistance in horseweed can be identified quickly with a leaf dip assay and plant growth stage will affect the impact glyphosate has on susceptible horseweed plants.

HAYFIELD SWARD RESPONSE TO APPLICATION TIMING OF AUXIN HERBICIDES WITH AND WITHOUT PENDIMETHALIN. A.P. Lassiter*, K.W. Bamber, and M.L. Flessner, Virginia Tech, Blacksburg, VA (8)

ABSTRACT

Buckhorn plantain (*Plantago lanceolate* L.) is a cool season perennial plant and is commonly encountered in pastures and hayfields throughout Virginia and most of the eastern U.S. This broadleaf weed increases in density over time and can compete with desired forage species, reducing yield. Due to its short growth habit, mowing is ineffective at controlling buckhorn plantain, necessitating herbicide treatment. The objective of this research was to evaluate buckhorn plantain control and sward response to auxin herbicides, with and without pendimethalin, applied in the spring versus fall.

A study was conducted in a buckhorn plantain infested hayfield in Brunswick County, Virginia from the fall of 2017 into the summer of 2018. Treatments were arranged in a randomized complete block design with four replications. Fall treatments were applied when buckhorn plantain was approximately 15 cm tall (September 13, 2017) and spring treatments were applied when buckhorn plantain was between 8 and 10 cm tall (April 13, 2018). Data collected included visible cover by species and buckhorn plantain control taken on April 13, 2018 and on June 6, 2018. Treatments were rated on a 0 (no control/cover) to 100 (complete control/cover) scale. Winter annual weed cover data (consisting of common chickweed (*Stellaria media*), Persian speedwell (*Veronica persica*), and hairy bittercress (*Cardamine hirsuta*)) were combined for analyses. Similarly, annual grass cover (consisting predominately of large crabgrass (*Digitaria sanguinalis*) and barnyardgrass (*Echinochloa crus-galli*)) were combined. Data were analyzed in JMP Pro 13 using ANOVA with subsequent means separation using Fisher's Protected LSD at $\alpha = 0.05$. Additionally, contrasts were conducted in order to determine the effect of pendimethalin addition as well as application timing.

2,4-D at 1.1 kg ha^{-1} + dicamba at 0.56 kg ha^{-1} with or without pendimethalin at 2.24 kg ha^{-1} resulted in 94 and 89% buckhorn plantain control, respectively, on April 13thwith no differences observed between treatments. Tall fescue cover was greatest (46%) from 2,4-D + dicamba + pendimethalin when evaluated on April 13, despite an increase in winter annual weed cover (23%) relative to the nontreated check (11%). 2,4-D + dicamba (without pendimethalin) resulted in greater fescue cover (27%) than the nontreated check (5%) but resulted in the greatest winter annual weed cover (42%) on April 13th. When evaluated in June, 2,4-D + dicamba + pendimethalin containing treatments and triclopyr at 0.56 kg ha^{-1} + 2,4-D + pendimethalin reduced weed cover (buckhorn plantain + annual grasses) the greatest ($\leq$21%) and had the most fescue cover (81 to 94%). Contrasts indicate that the addition of pendimethalin resulted in increased fescue cover in most cases and that 2,4-D + dicamba + pendimethalin resulted in more fescue cover when fall applied versus spring. While 2,4-D + dicamba and triclopyr + 2,4-D both resulted in acceptable buckhorn plantain control, the addition of pendimethalin is needed to prevent annual weeds from filling voids and allowed fescue cover to increase. Future research should repeat this study to corroborate results. Also, forage yield should be collected.

EVALUATING WEED CONTROL AND CROP RESPONSE TO PREEMERGENCE HERBICIDE OPTIONS IN TIGERNUT (*CYPERUS ESCULENTUS*). L. Black*, B.L. Carr, A. Ayeni, and T.E. Besançon, Rutgers University, New Brunswick, NJ (9)

ABSTRACT

Tigernut or 'chufa' (*Cyperus esculentus* var. *sativus*), a 'relative' of the weedy yellow nutsedge, is gaining popularity in the United States as a high energy tuber crop known for the sweet and chewy taste of the tuber which has 40-45% gluten free carbohydrate, high fiber content, healthful fat profile, high oleic acid, and high phosphorus, potassium and Vitamins C & E. This poster highlights the results of a field experiment carried out in 2018 at Rutgers Horticulture Farm 3 in East Brunswick, New Jersey, to evaluate weed control and the response of two tigernut selections (V1 and V3 from West Africa) to seven preemergence herbicide treatments, namely: *S*-metolachlor (Dual Magnum) at 1.43 lb ai/acre, oxyfluorfen (Goal Tender) at 0.25 lb ai/acre, bensulide (Prefar) at 6 lb ai/acre, pendimethalin (Prowl H20) at 1 lb ai/acre, oxyfluorfen + bensulide at 0.25 +6 lb ai/acre, oxyfluorfen + pendimethalin at 0,25 + 1 lb ai/acre and oxyfluorfen + DCPA (Dacthal) at 0.25 + 6 lb ai/acre. Weedy and hand weeded checks were included in the treatments for comparison. Tigernut was grown as sprigs (transplants) on raised seedbed under black plastic mulch in a sandy loam soil with pH 6.8 and 1.8% organic matter. The soil received recommended 20-20-20 NPK fertilizer during land preparation. Herbicides were applied with a 2-nozzle CO_2 backpack sprayer to a carefully cut out 4x1 ft^2 area at the top of the seedbed where tigernut sprigs were transplanted (at 12"spacing between sprigs) to allow proper evaluation of herbicide and weed impact. Results showed there was no significant difference in the response of tigernut selections to weed control treatments; *S*-metolachlor significantly inhibited tigernut growth and tuber yield (1,190 lb/acre) while oxyfluorfen + pendimethalin produced the highest growth and tuber yield (4,130 lb/acre). Except for *S*-metolachlor and bensulide alone (1,920 lb/acre) or combined with oxyfluorfen (2,720 lb/acre), all herbicide treatments produced tigernut tuber yield comparable to the hand weeded check (3,270 lb/acre). All herbicide treatments suppressed hairy galinsoga (*Galinsoga quadriradiata*), the dominant weed at the experimental site, satisfactorily up to 3WAP, but only oxyfluorfen + pendimethalin maintained acceptable control till 8WAP.

COMMON RAGWEED (*AMBROSIA ARTEMISIIFOLIA*) SURVIVAL AND FECUNDITY AFTER VARIOUS HERBICIDE TREATMENTS AND APPLICATION TIMINGS. E.B. Scruggs* and M.L. Flessner, Virginia Tech, Blacksburg, VA (10)

ABSTRACT

Common ragweed (*Ambrosia artemisiifolia* L.) is a troublesome weed for soybean growers due to its resistance to glyphosate and ALS inhibiting herbicides. Common ragweed (CR) control with herbicides is further limited by PPO resistance, recently confirmed.

Two field studies evaluated herbicides for CR control with an overarching goal of mitigating PPO resistance. Both studies were conducted in Brunswick County, Virginia in 2018. Both studies utilized a randomized complete block design with four replications and a nontreated check. All herbicide rates remained the same throughout both studies. Treatments in study one consisted of flumioxazin (89 g ai ha^{-1}), sulfentrazone (280 g ai ha^{-1}), or fomesafen (420 g ai ha^{-1}) applied PRE alone and followed by (fb) either mesotrione (105 g ai ha^{-1}) + glyphosate (1260 g ae ha^{-1}), dicamba (560 g ae ha^{-1}) + glyphosate, 2,4-D choline (1060 g ai ha^{-1}) + glyphosate, glufosinate (656 g ai ha^{-1}) + glyphosate, or fomesafen + glyphosate applied POST. Additionally, glyphosate was applied alone POST. PRE treatments were applied on May 22, 2018. All treatments included adjuvants and utilized nozzles as noted on product labels. POST treatments were applied on July 7, 2018, when weeds were ~15cm tall. Emerged weeds were flagged prior to POST application. Visible control was evaluated on a 0 (no control) to 100% (complete necrosis) scale 44 d after PRE. Both whole plot and surviving flagged weed count data were collected at POST and October 23, 2018 (mimicking soybean harvest). Treatments in study two included the following alone and in combinations with fomesafen: mesotrione, dicamba, 2,4-D choline, glufosinate, glyphosate, and fomesafen (alone only). Emerged weeds were counted and flagged according to size (5 to 10, 10 to 20, and 20 to 30cm tall) prior to application. POST treatments were applied on May 31, 2018. Count data of surviving flagged weeds by size class were collected on October 23, 2018 (mimicking soybean harvest). Data were subjected to ANOVA and subsequent means separation using Fisher's Protected LSD (α=0.05) using JMP Pro 14. To convert count data into control % data, counts of weeds surviving on October 23 were divided by the number of weeds initially flagged and subsequently subtracted from 1.

In study one, both flumioxazin and fomesafen resulted in >98% CR control 44 d after PRE. All treatments resulted in greater control than glyphosate (30%) alone. All dicamba containing treatments provided ≥96% control. On October 23 there were no differences in weed control due to PRE used. Flumioxazin fb 2,4-D or mesotrione did not result in commercially acceptable control of weeds (50%). In study two, 2,4-D alone and glufosinate alone both resulted in ≥94% control across size classes. Dicamba alone resulted in ≥88% control on 5 to 20cm weeds. Both dicamba and fomesafen alone resulted in 75% control of 20 to 30cm weeds. Mesotrione alone resulted in 50% and 6% control, respectively, on 10 to 20 and 20 to 30cm CR. Glyphosate performed poorly (≤25%) on all size classes. When combined with fomesafen, both glyphosate and mesotrione had improved control on all size classes (100%, 100%, and >75%, respectively). These studies reinforce the importance of targeting small (<10cm) weeds and indicate high efficacy of PPO containing PRE herbicides fb 2,4-D, dicamba, glufosinate, and fomesafen on CR. Future research should investigate the efficacy of these treatments on PPO resistant CR populations and other weed species.

GROW, A SCIENCE BASED RESOURCE TOOL FOR INTEGRATED WEED MANAGEMENT. C.G. Rubione*, K.B. Pittman, S. Mirsky, M.L. Flessner, L. Lazaro, M.J. VanGessel, and M. Bagavathiannan, University of Delaware, Georgetown, DE (16)

ABSTRACT

Weed management has always been challenging, but it has been complicated with the dramatic rate at which weed resistant species are developing, with 161 herbicide-resistant species reported in the US. Increasingly, growers, crop consultants, and other agricultural professionals are looking towards integrated weed management (IWM) to more effectively target herbicide-resistant weeds on their farms. IWM allows growers to use multiple tactics (chemical, cultural, mechanical, biological, and prevention) to attack weeds from multiple angles.

GROW, is a web-based resource for IWM providing science-based information to control herbicide-resistant weeds as well as other problematic weeds. Understanding weed behavior is a key starting point. This website provides users with information on weed biology and ecology for key weed species, as well as Apps for identification and scouting material. Website users will have direct access to a variety of resources to develop integrated weed management programs for their specific needs.

Apps for an effective site of action and Palmer Amaranth Management (PAM) are new tools being developed for GROW. State extension websites, newsletters, Ag journals, scientific papers, and a broad range of tactics adapted to various regions of the country are to be provided in the "Weed Management Toolbox" tab.

GROW's main objective is to promote and increase the adoption of IWM practices. It is committed to providing resources to help farmers and crop advisors to develop their own sustainable weed management practices.

GROW is a first step towards "Getting Rid of Weeds" through Integrated Weed Management.

http://integratedweedmanagement.org/

ESTABLISHING A WEED IDENTIFICATION NETWORK IN NEW YORK: IMPROVING EARLY DETECTION IN AN ERA OF RAPID WEED RANGE EXPANSION. C. Marschner* and A. DiTommaso, Cornell University, Ithaca, NY (17)

ABSTRACT

The Weed Management and Ecology lab at Cornell University received a grant in 2018 to establish a stronger network of agricultural weed identification services and outreach tools for New York. From 2018-2021, the lab will work with extension educators and crop advisors to identify weed identification knowledge gaps in New York grower communities, and provide trainings and targeted outreach tools to fill those gaps. We will also work to establish a method for the agricultural community to report priority and newly arrived weed species and biotypes, so that state agencies and growers can respond effectively to minimize impact on yields and profit.

EXPLORING THE LINK BETWEEN THE SOIL MICROBIAL COMMUNITY AND DIFFERENTIAL WEED SPECIES RESPONSE TO HIGH CARBON SOIL AMENDMENTS. M.A. Gannett*, A. DiTommaso, and J. Kao-Kniffin, Cornell University, Ithaca, NY (19)

ABSTRACT

Broad applications of nitrogen (N) fertilizers will often influence plant population dynamics that favor the establishment and expansion of nitrophilous (N-responsive) weed populations. Therefore, many agricultural weeds can out-compete crops in fields with high N availabilities. One way to reduce N availability to plants is through enhancing microbial immobilization processes in soil by incorporating carbon (C) amendments to stimulate microbial growth. Our study shows decreased growth of weed species when a C source is added to the soil. Some species were affected more than others, but this does not seem to be dictated by the root:shoot ratio. Since soil microbial growth was consistently stimulated with high C soil amendments, the microbial community and how it interacts with each plant may more strongly predict competitive advantage in high C soil environments. We will now test whether mycorrhizal or rhizobial associations with soybean can improve crop/weed competition in high C soil environments compared to unamended environments.

USE OF DRONE IMAGING FOR ASSESSING WEED CONTROL AND DISEASE PRESSURE IN HIGHBUSH BLUEBERRY. M.G. Mars, D.C. Nuhn, B.L. Carr*, T.E. Besançon, and P.V. Oudemans, Rutgers University, Chatsworth, NJ (23)

ABSTRACT

New Jersey, the Garden State, is home to the largest diversity of specialty crops in the North East region. One in particular, the blueberry, is a major cash crop for which New Jersey ranks 5th in the United States. Understandably, weed detection and prevention remains crucial in order to prevent crop loss. In recent years, the technology behind unmanned aerial vehicles (UAVs) has jumped forward by leaps and bounds, allowing for cost-efficient, high-resolution, real-time spatial data collection and analysis. This study was conducted to evaluate the efficacy of drone imagery for weed detection and growth over time in highbush blueberry crops. The study focused on two rows of cultivar Duke blueberries. A study including various preemergence (PRE) herbicide treatments and a weedy check was set up in these two rows and plots were monitored throughout the growing season. The efficiency of small unmanned aerial vehicle (sUAS) detection was compared throughout the growing season with other detection methods, namely, the Canopeo application (http://www.canopeoapp.com/) and manual weed density counts. Two sensors were utilized for sUAS data collection: a broad band RGB (Red Green Blue) sensor and a narrow band multispectral (Blue, Green, Red, Red-Edge, and Near Infrared) sensor. The sensors (especially the multispectral one), though very useful for image capture and general field analysis, are limited by their resolutions. One issue that cropped up was being unable to differentiate between weeds and the edges of the bushes, as they had very similar reflectance values. The results showed high correlation levels between the Canopeo data and the RGB-based Green Leaf Index (GLI) data ($R^2 = 0.82$), and high correlation levels between the GLI data and the multispectral NDVI (Normative Difference Vegetative Index) data ($R^2 = 0.96$). The results also saw poor correlation levels between sUAS-gathered GLI data and weed density data (R^2=0.59), and between Canopeo data and weed density data (R^2=0.53). It is clear from both the RGB and multispectral imagery that the different herbicide PRE treatments expressed various degrees of weed control success, ranging from less than 20% control 35 weeks after treatment (WAT) with dichlobenil to 100% control 18 WAP with a combination of diuron, mesotrione, and oryzalin. However, additional studies are warranted to determine if some of the treatments tested may have negative effects on crop development, given the predominantly sandy texture of New Jersey soils. Future research will also be directed toward the use of machine learning for improving disease and weed tracking through the use of drone imagery.

CRANBERRY RESPONSE TO RATE AND APPLICATION TIMING WITH FLUMIOXAZIN AND SULFENTRAZONE. B.L. Carr* and T.E. Besançon, Rutgers University, Chatsworth, NJ (24)

ABSTRACT

In 2017, New Jersey cranberry production estimated 29.6 million kg of cranberries at a farm value of $28 million (USDA 2017). New Jersey cranberry production is concentrated in the Pine Barrens coastal plain where sandy acidic soils are optimal for cranberry. Due to the specific edaphic conditions and perennial nature of cranberry production, weed control remains a major challenge. Therefore, through the IR-4 program, it is important to screen and evaluate herbicides already registered on other minor crops that can provide effective control of though-to-control weeds in cranberry production. Field studies were conducted in 2018 to test weed control efficacy and crop tolerance of two non-registered herbicides, sulfentrazone and flumioxazin. Trials were laid out in a 'DeMoranville' cranberry bog at the Rutgers P.E. Marucci Center in Chatsworth, NJ, as wells as in an 'Early Black' cranberry bed at Theodore H. Budd and Sons Farm in Southampton, NJ. Similar studies were also conducted in Wisconsin and Massachusetts. Treatments consisted of two different rates for each herbicide (flumioxazin at 143 and 286 g a.i. ha^{-1} and sulfentrazone at 280 and 420 g a.i. ha^{-1}) applied in spring when crop is still dormant at two different timings: early (bud swell cranberry stage) and late (cabbage head cranberry stages). A non-treated check (UNT) was included in each study. Necrosis of the terminal bud and cranberry stunting were rated 2, 4, and 8 weeks after treatment (WAT). Shoots were evaluated for primary and axillary shoot number, length, and number of fruits 8 WAT and at harvest. Yield was measured at harvest. Flumioxazin at 286 g a.i. ha^{-1} applied at the cabbage head stage caused the highest amount of terminal bud necrosis 2 and 4 WAT (60% and 14%, respectively). Flumioxazin at 143 a.i. ha^{-1} at bud swell and sulfentrazone at 280 g a.i. ha^{-1} at cabbage head stage caused less than 5% necrosis 2 WAT. By 4 WAT, necrosis symptoms virtually disappeared for sulfentrazone at 280 g a.i. ha^{-1} applied at bud swell. Flumioxazin at 286 g a.i. ha^{-1} applied at the stage resulted in the highest level of crop stunting 2 and 4 WAT (96 and 51%, respectively). Sulfentrazone at 280 g a.i. ha^{-1}applied at bud swell resulted in the lowest amount of stunting 2 and 4 WAT (41 and 8%, respectively). Overall, yield was extremely low for the DeMoranville cultivar, regardless of treatments. In the Early Black trial, sulfentrazone at 280 g a.i. ha^{-1} at bud swell did not significantly decrease fruit yield compared to UNT. Flumioxazin, regardless of application rate or timing, caused at least 90% yield reduction compared to UNT. Development of non-productive axillary shoots was multiplied by 3 in average with late application of sulfentrazone at 280 g a.i. ha^{-1} or flumioxazin, regardless of rate, for the DeMoranville cultivar.

CONTROL OF BITTERCRESS AND OXALIS WITH PREEMERGENCE HERBICIDES OVER TIME.
J. Altland*, USDA-ARS, Wooster, OH (25)

ABSTRACT

[no abstract submitted]

PREEMERGENCE HERBICIDE SAFETY TO SEDUM RUPESTRE AND SEDUM ACRE. C. Harlow* and J. Neal, North Carolina State University, Raleigh, NC (26)

ABSTRACT

Angelina stonecrop, *Sedum rupestre* 'Angelina', and Goldmoss stonecrop, *Sedum acre* 'Aurea', are evergreen low-growing groundcovers that are drought tolerant. Sedums are relatively easy to propagate but do not have deep root systems and do not tolerate competition from weeds. Few herbicides are labeled for weed control in sedum production.

Experiments were initiated to evaluate the safety of preemergence herbicides for use in container-grown Angelina stonecrop and Goldmoss stonecrop. Herbicides evaluated on Angelina stonecrop were dithiopyr (Dimension EW) at 2, 4 or 8 pt/A; oxadiazon (Ronstar G) at 100, 200 or 400 lb/A; prodiamine + isoxaben (Gemini G) at 200, 400 or 800 lb/A; and pendimethalin (Pendulum 2G) at 150, 300 or 600 lb/A. Dimethenamid + pendimethalin (Freehand) at 150 lb/A is currently labeled for use in sedum production and was included for comparison. Prodiamine + isoxaben (Gemini G) at 200, 400 or 800 lb/A was evaluated on Goldmoss stonecrop. Dimethenamid + pendimethalin (Freehand) at 150 lb/A and pendimethalin (Pendulum 2G) at 150 lb/A were included for comparison. Rooted liners of both species were potted into 3-L pots on May 9, 2018 and hand-watered to settle the substrate. Additionally, cuttings of Angelina stonecrop were placed on the surface of the substrate and lightly covered with additional bark substrate. These unrooted pots were treated with the low (1X) rate of Ronstar G, Gemini G, Pendulum 2G or Freehand. The experiments were conducted in RCBD with 4 replicates. Herbicides were applied on May 9, 2018 and again 6 weeks later on June 19, 2018. Plant injury was visually evaluated on a 0 to 10 scale where 0 = no visible injury and 10 = dead plants. Additionally, plant canopy volume was estimated 6 weeks after the 1st treatment application (WAT) for *Sedum acre*, and crop cover (% area of pot surface covered) was measured 2 weeks after the 2nd treatment application (WAT2) for both species.

Severe injury to both *Sedum* species was observed on plants treated with any dose of Gemini G or Freehand at 150 lb/A. Up to 25% injury to both *Sedum* species was observed on plants treated with Pendulum 2G at 150 lb/A, with greater injury to *S. rupestre* at higher doses. Severe injury to *S. rupestre* was observed on plants treated with Dimension EW at any dose. *S. rupestre* treated with Ronstar showed no injury at 100 lb/A; however, up to 15% injury was observed in plants treated with 200 or 400 lb/A. Results for unrooted cuttings of Angelina stonecrop treated with the lowest rate of Ronstar G or Pendulum 2G were similar to transplants. More severe injury was observed on unrooted cuttings treated with Gemini G or Freehand compared to transplanted Angelina stonecrop.

HOW WILL EXTENDED DROUGHTS AFFECT BURCUCUMBER (*SICYOS ANGULATUS*) COMPETITION IN SILAGE CORN IN NEW YORK? A. DiTommaso*, K.M. Averill, S.H. Morris, and M.C. Hunter, Cornell University, Ithaca, NY (27)

ABSTRACT

Climate change is expected to affect the Northeastern United States with extended droughts during the growing season. However, little is known about how weed species will respond to these altered environmental conditions. Burcucumber is a weedy annual vine that is currently challenging to manage in the Southeastern and the Mid-Atlantic US and, based on surveys of growers and extension agents, is increasing in NY. In Ithaca, NY in 2018, a drought and competition experiment was established to assess the impact of burcucumber on corn (*Zea mays* L.) silage yield. Corn was planted on May 29 and 10 days after planting (DAP), burcucumber seedlings (of similar size to corn seedlings, both ~15 cm) were transplanted into corn rows at four densities (0, 0.5, 2, 3 plants m^{-2}). Other weeds were controlled with inter-row cultivation 23 DAP and weekly hand-weeding. At 24 DAP, a 38-day drought treatment was imposed using rainout shelters, 2.7 m wide by 3 m long by 3 m tall, constructed with high clarity plastic and gutters and drainage tubing to divert water away from drought plots. Soil water availability was estimated at least once per week during the experiment using fixed-in-place gypsum block electrodes. Corn silage and burcucumber biomass were harvested on September 14 (108 DAP). Mixed model ANOVA and regression were conducted with drought, burcucumber density, and their interaction as fixed effects and block and corn density as random effects. Estimated soil water availability was lower in drought plots (47 ± 1) compared with non-drought plots (69 ± 1) ($P < 0.05$). Drought plots received 56% of the growing season rain (24 cm) relative to non-drought plots (43 cm) (May 1 to harvest). Results showed that burcucumber ($P < 0.05$) and drought ($P < 0.05$) reduced corn silage yield; the interaction was not significant ($P > 0.05$). Fresh corn silage yield was reduced by 29% in the drought treatment (17 ± 7 Mg ha^{-1}) relative to the non-drought control (24 ± 9 Mg ha^{-1}). Yield was reduced by 23% from 24 ± 3 Mg ha^{-1} with no burcucumber to 18 ± 3 Mg ha^{-1} with three burcucumber plants m^{-2}. Higher burcucumber densities and a longer drought than tested in our study would be expected to reduce yield further. Based on the first year of this experiment, burcucumber and an extended drought independently have the capacity to reduce corn silage yield.

WSSA ADVOCATES FOR WEED CONTROLS THAT PROTECT SOYBEAN EXPORT VALUE.

C.M. Moseley*, L. Van Wychen, H. Curlett, J. Schroeder, P.D. Laird, and S.P. Conley, Syngenta, Greensboro, NC (28)

ABSTRACT

Weeds and weed seeds are a serious phytosanitary concern. Most countries, including the United States, take action when weed seeds are detected in arriving shipments. The importing country may reject, re-export, or destroy the shipment. In the worst case, the country may suspend imports or close the market altogether. Soybeans are one of the United States' top exports. Increases in herbicide-resistant weeds may be contributing to more weed seeds in harvested beans. There are a number of best practices—many of which are already in use here in the United States—that can be applied on farm and by grain handlers to help reduce weed seeds in U.S. soybeans. These practices are being promoted by USDA, WSSA, and other organizations across the country.

JAPANESE STILTGRASS (*MICROSTEGIUM VIMINEUM*) MANAGEMENT IN A PASTURE: USING A GDD MODEL TO APPLY PENDIMETHALIN. R.S. Chandran*, D. Shipman, and B.M. Loyd, West Virginia University, Morgantown, WV (29)

ABSTRACT

Japanese stiltgrass (*Microstegium vimineum*) is an unpalatable invasive annual grass with limited selective management options in a cool-season pasture. Pendimethalin is considered to be an effective herbicide to manage this weed when applied prior to germination. We hypothesized that a predictive model to time pre-emergent application based on a Growing Degree Days (GDD) may provide consistent weed control. An experiment was conducted in 2018 at Bethlehem, West Virginia, to time pendimethalin application at 4.48, 2.24, and 1.12 kg ha^{-1}, based on a GDD Model available through Climate Smart Farming at Cornell University (http://climatesmartfarming.org/) prior to weed germination. Herbicide treatments were applied on April 27, when GDD_{10C} recorded 134 at this location. Japanese stiltgrass germinated a week later in untreated plots when GDD_{10C} surpassed 200. Soil moisture levels preceding and following treatments were adequate to ensure herbicide activation. Pendimethalin applied at 4.48 kg ha^{-1}consistently provided excellent control (>95%) of Japanese stiltgrass up to 16 weeks after treatment (WAT), whereas pendimethalin applied at 2.24 and 1.12 kg ha^{-1} failed to provide acceptable levels of control (<70%) during this time. A GDD model, by virtue of its direct relationship with soil temperature (warming), could be used as a viable estimate to time effective preemergence herbicide application to manage Japanese stiltgrass in a cool-season pasture upon validation.

ROLE OF N-FERTILIZATION AND HERBICIDES IN INTEGRATED MANAGEMENT OF MUGWORT (*ARTEMISIA VULGARIS* L.) IN A COOL-SEASON GRASS PASTURE. J.S. Aulakh*, The Connecticut Agricultural Experiment Station, Windsor, CT (30)

ABSTRACT

Mugwort has invaded diverse types of habitats comprising roadsides, floodplains and riparian areas, pastures and rangelands, and various agronomic, turf, and landscape settings. A three year field experiment was initiated in the fall of 2015 at the Lockwood Research Farm of the Connecticut Agricultural Experiment Station in Hamden, CT for mugwort management in a cool-season grass pasture. The experimental design was a split-split plot with three replications. The treatment design was a 3^3 factorial of three nitrogen levels (0, 62, and 124 kg ha^{-1}) in the main plot, three herbicides (aminopyralid, clopyralid, and glyphosate) in the sub plot, and three application rates in the sub-sub plot. The application rates were: 0.06, 0.12, and 0.24 kg ae ha^{-1} of aminopyralid, 0.14, 0.28, and 0.56 kg ae ha^{-1} of clopyralid, and 0.55, 1.1, and 2.2 kg ae ha^{-1} of glyphosate. Each year the entire experiment site was mowed in early August followed by application of nitrogen as fertilizer urea (46-0-0) in late-August or early September. Herbicide treatments were applied in October with a compressed CO_2 sprayer fitted with four XR Teejet 8002 nozzles delivering 20 GPA at 40 psi. Data analyses revealed herbicide by application rate by months-after-initial-treatment (MAIT) interactions for percent visual control and rhizome dry weight, and a nitrogen by herbicide by application rate interaction for dry weight of cool-season grasses. Aminopyralid at $\geq$ 0.12 kg ae ha^{-1} and glyphosate at $\geq$ 1.1 kg ae ha^{-1} were highly effective with a visual mugwort control of 90 to 99% over the three years. Mugwort rhizome dry weight also showed similar reduction compared to the nontreated control (no nitrogen and no herbicide). Clopyralid control of mugwort was not satisfactory. Cool-season grass dry weight was highest in the aminopyralid treated plots that received $\geq$ 62 kg ha^{-1} of nitrogen due both to the higher mugwort control and improved N-fertilization. Clopyralid or glyphosate treated plots did not show differences in grass dry weight from improved nitrogen due to the lack of mugwort control by clopyralid and removal of cool season grasses by glyphosate.

UNDERSTANDING THE INTERACTION OF CROP MANAGEMENT SYSTEM AND CROP IDENTITY ON BIOLOGICALLY-MEDIATED PLANT-SOIL FEEDBACKS. U.D. Menalled*, T. Seipel, and F.D. Menalled, Cornell University, Ithaca, NY (31)

ABSTRACT

Cropping system characteristics such as tillage intensity, cover cropping, and the application of off-farm synthetic inputs influence weed abundance and crop-weed competition. The resulting plant community has species-specific effects on soil microbial communities which, impact the emergence, growth, and competitive ability of subsequent plants and create a plant-soil feedback (PSF) loop. This study assessed the impact of crop identity and 1) a chemical no-till system, 2) an USDA-certified organic system reliant on tillage and 3) an USDA-certified grazed organic system with reduced tillage on crop emergence and PSFs. The three farming systems followed a five-year rotation of safflower (*Carthamus tinctorius*) inter-seeded with yellow sweet clover (*Melilotus officinalis*), yellow sweet clover as cover crop, winter wheat (*Triticum aestivum*), lentil (*Lens culinaris*), and winter wheat. Soil samples were collected in the safflower, yellow sweet clover, and year three winter wheat plots. Sterile greenhouse soil was inoculated with 4%, by volume, field soil and crops growing during the sample year were grown twice for five weeks in a greenhouse to condition soils with crop-specific microbial communities. The following crop in the field rotation was then grown for a seven-week response phase. This procedure was replicated in uninoculated soil. PSFs were calculated by comparing the biomass of response plants grown in the inoculated and uninoculated soil. Emergence did not differ across systems but was affected by crop identity. Organic reduced-till systems harbored the most positive mean PSFs and organic till systems the most negative PSFs. Chemical no-till system PSFs were intermediate between those of the two organic systems. Differences between organic system PSFs suggest that the combined effect of tillage and livestock incorporation may impart a larger effect on PSFs than chemical management. A farm system and crop identity interaction suggests that farm management has species-specific PSF effects.

INTERACTION OF ORGANIC HERBICIDES AND COVER CROPS ON WEED SUPPRESSION AND INSECT DYNAMICS IN NO TILL SYSTEMS. D.G. Lewis*, M.A. Cutulle, R. Schmidt-Jeffris, C. Blubaugh, T. Campbell, and J. Coffey, Clemson University, Clemson, SC (32)

ABSTRACT

Cover crops have been shown to improve many aspects of agroecosystems; they can suppress weeds, capture run-off, improve soil health, and increase beneficial insect biodiversity and weed seed biological control. Incorporating organic herbicide applications with cover crops could help in achieving optimal weed control throughout the season and is something that has not been studied. The purpose of this 2-year project is to determine which cover crop and herbicide program result in the best pest control of arthropods and weeds. The impacts of cover crops and organic herbicides on arthropod pest pressure and natural enemies were quantified by various traps such as fire ant vials and pitfall traps within each plot. We also monitored weed pressure and weed seed biological control. We seek to investigate how these combinations will achieve maximum integrated control of weeds and insect pests.

IMPLICATIONS OF A BIPHASIC MODEL TO PREDICT WILD RADISH (*RAPHANUS RAPHANISTRUM* L.) ERGENCEEE. T.A. Reinhardt Piskackova*, C. Reburg-Horton, R.J. Richardson, K.M. Jennings, and R.G. Leon, North Carolina State University, Raleigh, NC (33)

ABSTRACT

Wild radish emergence has been difficult to model due to irregular emergence patterns, which limit the applicability of traditional single models including log-logistic, Gompertz, and Weibull. The emergence of wild radish was studied in agricultural fields in Clayton, NC from September to June. Chronological, Thermal, and Hydrothermal time were used to describe seedling emergence dynamics. Even with temperature and moisture restrictions, wild radish demonstrated discontinuous emergence that was best described as a biphasic pattern. For this reason, a hybrid model combining sigmoidal and Weibull equations was developed to fit the observed emergence. This type of biphasic pattern may be a survival strategy for wild radish related to low temperatures during the winter.

TEMPERATURES DURING AND AFTER TREATMENT AFFECT ZOYSIAGRASS RESPONSE TO GLYPHOSATE AND GLUFOSINATE. J.M. Craft*, S.D. Askew, and J. Brewer, Virginia Tech, Blacksburg, VA (34)

ABSTRACT

Turf managers typically exploit winter dormancy of warm-season grasses (e.g., bermudagrass, *Cynodon* spp.) to utilize non-selective herbicides. Glyphosate and glufosinate are labeled for use in dormant zoysiagrass but turf specialist rarely recommend this practice for fear that zoysiagrass may not be "completely dormant." Research has shown that air temperatures can affect weed control and crop safety from herbicides and we hypothesized this to be the case with zoysiagrass. However, no research has examined the impact of temperature during or after non-selective herbicide applications on zoysiagrass response during spring green-up. The objective of this research was to determine the influence of three different temperature regimes during and after application on zoysiagrass response to glyphosate and glufosinate.

A randomized complete block study was initiated that included 6 replications in time. Treatments included a split plot arrangement of 3 temperature main plots and a 2 (herbicides) x 3 (rates) factorial arrangement of subplots. Dormant 'Meyer' zoysiagrass plugs were greenhouse incubated for 48 hours on March 6, 2018. The plugs were dissected and five, 5-cm sprigs that exhibited early development of green foliage were placed in petri dishes on moist paper discs. Petri-dishes were then placed into three growth chambers set to constant 10°, 18°, and 27° C. After 12 hours, petri-dishes were removed from growth chambers and over sprayed with glyphosate at 75, 350, or 700 g ai ha or glufosinate at 420, 840, or 1680 g ai ha then returned to chambers. A nontreated check was included in each temperature regime. Seven days after treatment (DAT), petri-dishes were moved to a green house at 27°C and incubated an additional 14 days. During incubation, zoysiagrass green cover was assessed 11 times via aerial digital images that were analyzed to enumerate green pixels. At 21 DAT, sprigs were dried for 48 hours and weighed. Detected green pixels of treated sprigs were converted to a percentage reduction relative to nontreated sprigs for each temperature by replicate combination. Percent reduction in green pixel count over time was subjected to hyperbolic or sigmoidal nonlinear regression to determine the number of days required to reduce growth by 50% (GR50). A unique GR50 value was determined for each experimental unit and these data were subjected to ANOVA and Fisher's Protected LSD separation. Dry biomass of sprigs was subjected to similar ANOVA and mean separation.

Data indicated that zoysiagrass sprigs treated with glufosinate reduced GR50 more rapidly than glyphosate regardless of rate. Zoysiagrass incubated for 7 days at 10° C required 9 days to reach a herbicide-induced GR50 and significantly more time than zoysiagrass incubated at 18° C and 26° C. Glufosinate reduced 26° C incubated zoysiagrass growth by 50% in less than one day and faster than other treatments. Glyphosate had essentially no influence on 10° C incubated zoysiagrass growth until after sprigs had been moved to higher temperatures 7 DAT and GR50 values typically ranged between 7 and 11 days. Results indicate as temperatures increase the rate of glyphosate and glufosinate activity increases. These data suggest that glyphosate activity on zoysiagrass is considerably more temperature dependent than that of glufosinate.

ANNUAL BLUEGRASS WEEVIL (*LISTRONOTUS MACULICOLLIS*) AND PACLOBUTRAZOL FOR ANNUAL BLUEGRASS (*POA ANNUA*) CONTROL IN FAIRWAYS. K. Diehl*, M. Elmore, A. Koppenhöfer, J. Murphy, D.P. Tuck, and O. Kostromytska, Rutgers University, New Brunswick, NJ (35)

ABSTRACT

The annual bluegrass weevil (*Listronotus maculicollis*; ABW) is a common insect pest of cool-season turfgrass. The ABW showcases ovipositional preference towards annual bluegrass (*Poa annua*; ABG) over creeping bentgrass (*Agrostis stolonifera*).

Research was conducted in 2018 at the Rutgers Horticulture Farm No. 2 in North Brunswick, NJ on a simulated creeping bentgrass and annual bluegrass fairway to investigate various insecticide and paclobutrazol programs for annual bluegrass control. Three insecticide programs and four rates of paclobutrazol were evaluated in a three-by-four complete factorial randomized complete block design with four replications. Paclobutrazol rates consisted of a high (140 g ha^{-1}), medium (70 g ha^{-1}), and low rate (50 g ha^{-1}), as well as a non-paclobutrazol-treated control. Paclobutrazol was applied monthly from 15 May to 19 October. The first insecticide program (preventative) was designed to prevent any turfgrass injury. The second program (threshold) consisted of an insecticide applied after visual ratings determined that ABW damage resulted in unacceptable (<6.0) turfgrass quality. The third program (no-insecticide) was not treated with insecticides for ABW control.

Percent annual bluegrass cover was evaluated visually each month and grid intersect counts were conducted in June and November. Turfgrass quality, annual bluegrass quality, and creeping bentgrass quality were evaluated monthly on a 1 (poor) to 9 (excellent) scale where 6 is considered acceptable. Lightbox photos were taken monthly and subjected to digital image analysis (Turf Analyzer software). ABW larvae were sampled on 4 June to determine the efficacy of each insecticide treatment. Data were subjected to ANOVA in SAS using the GLIMMix procedure and Fisher's Protected LSD (α=0.05) was used to separate means.

The main effects of insecticide program and paclobutrazol were significant for annual bluegrass cover throughout the growing season; in November the main effect interaction was significant. Only the highest order interaction results from November is discussed for brevity. Within each insecticide program, paclobutrazol reduced annual bluegrass cover compared to the non-paclobutrazol-treated control. Compared to the preventative insecticide program, the threshold and no-insecticide programs resulted in less annual bluegrass cover for the non-paclobutrazol-treated control and medium rate of paclobutrazol. In the absence of paclobutrazol, the preventative program resulted in greater annual bluegrass cover (82%) than the threshold and no-insecticide programs (67 and 56%, respectively). The threshold and no-insecticide programs treated with the medium paclobutrazol rate resulted in less annual bluegrass cover (13 and 8%, respectively), than the preventative program (33%). The high and low rates of paclobutrazol resulted in 7 to 14% and 34 to 37% annual bluegrass cover and were not affected by insecticide program. This experiment will be repeated in 2019.

EFFECT OF PLANTING DATES ON SOYBEAN RESPONSE TO DICAMBA. M.A. Granadino*, W.J. Everman, J.T. Sanders, D.J. Contreras, and E.A. Jones, North Carolina State University, Raleigh, NC (36)

ABSTRACT

In the last two decades, the use of dicamba has increased to control glyphosate-resistant weeds (Alves et al 2018). Dicamba-resistant crop varieties have the potential to become widely utilized as a tool to manage glyphosate-resistant weeds in North Carolina. Dicamba sprayed as low as 0.04 g ae h^{-1} has been reported to reduce yield by up to 10% (Weidenhamer et al 1989). Soybean has been found to be more susceptible to dicamba applications when exposed at flowering (R1 and R2) compared to vegetative stages (V1 to V7) (Kniss 2018). This research was conducted to evaluate injury of sub-lethal rates of dicamba in maturity group V and VI soybean cultivars at V4 and R2, and evaluate if visual injury is an accurate method of estimating yield loss. Two separate studies were conducted in 2018 in Kinston and Rocky Mount, NC. Study group was organized by soybean maturity group and planting timing (May and June). Plots consisted of two borders rows to minimize potential drift. Dicamba applications were made with a pressurized CO_2 backpack sprayer. Dicamba was sprayed at 0.44, 1.76 and 7.04 g ae a^{-1} at V4 and R2 at each respective planting date. Weekly visual injury ratings and height measurements as well as final yield were collected. At 7.06 g ae a-1 applied at V4, dicamba had a more significant effect on injury than applied at R2. There was no significant difference on injury at 0.44 and 1.76 g ae a-1 between application timing. Soybean that was planted in May and sprayed at R2 had no significant difference in injury during the whole season as soybean planted in June and sprayed at V4. Soybean planted in May all rates had a significant impact on yield, with no significant difference in yield reduction among applications. There was no significant impact on yield between application timings for soybeans planted in May, but there was a difference between application dates when planted in June. Application timing had a greater impact on yield on soybean planted in June and sprayed at R2. Applications at V4 have no significant impact on yield regardless of planting date. Planting date has no significant effect on injury. Visual injury is not an accurate method of estimating yield loss.

TRICLOPYR AMINE BASAL BARK TREATMENTS CAN REPLACE OIL BASED STEM TREATMENTS IN AQUATIC SETTINGS. E. Weaver* and A. Gover, Penn State University, State College, PA (37)

ABSTRACT

For control of large invasive trees and shrubs, stem treatments are highly useful because they require less chemical, prevent sprouting, and reduce non-target injury. Common woody stem treatments include hack-and-squirt and basal bark. Hack-and-squirt requires applying herbicide into spaced cuts into the vascular tissue around the entire stem. In the basal bark treatment an oil-soluble herbicide is mixed with an oil carrier and sprayed on the entire surface of the lower 25 to 50 cm of the stem. Current basal bark herbicides are not aquatic-registered, limiting their use in aquatic areas. We conducted a series of experiments comparing water-soluble triclopyr treatments to a triclopyr-ester standard treating individual plants, 10 plants per treatment arranged in a Completely Randomized Design. Visual ratings of canopy reduction (injury) were subjected to analysis of variance, and means separated using Fisher's Protected L.S.D. In August 2015 we compared three experimental treatments to two standards and a control on European alder [*Alnus glutinosa* (L.) Gaertn.] at Bald Eagle State Park, Howard, PA. Standard treatments were triclopyr-amine hack-and-squirt at 0.14 g ae/cm diameter breast height (DBH), and triclopyr-ester basal bark at 0.071 g ae/cm DBH. The experimental treatments consisted of triclopyr-triethanolamine applied at 0.071, 0.11, or 0.14 g ae/cm DBH, diluted in a 1:1 mix of water and basal oil with emulsifier (Arborchem Basal Oil) that constituted 75, 62, and 50 percent of the mix by volume, respectively, with increasing triclopyr rate. When rated 11 months after treatment (MAT), alder injury was rated at 63, 86, and 92 percent for the 0.071, 0.11, and 0.14 g ae experimental treatments, respectively, and 100 and 90 percent injury for the hack-and-squirt and basal standards, respectively. On April 11, 2017, we applied triclopyr-choline at 144 g ae/L alone, or in combination with imazapyr at 4.8 g ae/L or aminopyralid at 2.4 g ae/L, or triclopyr-ester at 96 g ae/L as a standard to European alder and autumn olive (*Elaeagnus umbellata* Thunb.) at Canoe Creek State Park, Hollidaysburg, PA. The triclopyr-choline treatments were diluted in a 1:1 mix of an aquatic-registered crop oil concentrate (Agri-Dex) and water, while the triclopyr-ester treatment was diluted in basal oil. The treatments were applied to simulate an operational application, lightly covering the lower 30 cm of each stem, using a CO_2-powered, hand-held sprayer equipped with a single TeeJet #5500 Adjustable ConeJet with X-2 tip. In a separate experiment at the same location we applied triclopyr-choline or triclopyr-ester at 0.29 or 0.24 g ae/cm DBH, respectively to a separate stand of European alder on November 8, 2017 to test application timing effects. Treatments applied in April caused 90 to 100 percent canopy reduction to both species by 14 MAT and there was no significant difference between any herbicide treatments. Triclopyr-choline treatments applied in November caused significantly less canopy reduction than the triclopyr-ester (48 vs. 78 percent) 10 MAT. We believe water soluble triclopyr basal bark treatments are effective in the early spring and active growing season, and therefore provide a useful tool for vegetation managers working in settings near water.

CRITICAL PERIOD OF WEED CONTROL FOR GRASS SPECIES IN GRAIN SORGHUM. D.J. Contreras* and W.J. Everman, North Carolina State University, Raleigh, NC (38)

ABSTRACT

Grain sorghum is the 5th most planted crop in the world, as well as 4th in the US. Its production is mainly concentrated in the Midwest, where precipitation is limiting for high yielding corn and soybean production. Weed management is a concern for grain sorghum growers due the limited number of herbicide options available. Grass weed species have been able to thrive under conditions were most herbicide control programs are focused on broadleaf weeds. There are many options available for broadleaf weed control but very few exist for selective annual grass weed management. Some tools to consider for grass weed management are: the use of specialized varieties such as that containing the Inzen Sorghum trait (tolerance to nicosulfuron and rimsulfuron), the use of narrow rows, and the use of the Critical Period of Weed Control (CPWC) defined as a period in a crop´s growth cycle during which weeds must be controlled to prevent yield losses. The objective of this study was to determine the CPWC for grass species in a grain sorghum using a variety containing the Inzen™ Sorghum Trait and two different row width arrangements (15 and 36 in.) in two different locations in North Carolina (Rocky Mount and Lewiston). Plots were kept weedy or weed free through 2, 3, 5, and 7 weeks after planting. Two controls consisting of weedy and weed free all season plots were used for a treatment comparison based on a regression analysis where time of weed removal was related to crop yield. The CPWC to reduce yield losses to 5% or less for 15" rows at Rocky Mount was from 3.3 – 7.1 weeks after planting, and 1.3 – 6.7 weeks after planting for 36" rows. In Lewiston, the CPWC for 15" rows was from 2.4 – 8.9 weeks after planting, and 1.4 – 7.3 weeks after planting for 36" rows.

DETERMINING SORGOLEONE EFFICACY ON SELECT WHEAT (*TRITICUM AESTIVUM* SSP.) CULTIVARS AND WEED SPECIES. E.A. Jones*, M. Bansel, W.J. Everman, J.T. Sanders, D.J. Contreras, and M.A. Granadino, North Carolina State University, Raleigh, NC (39)

ABSTRACT

Sorghum-wheat rotations have been examined as a potential crop system for North Carolina. Sorghum produces an allelochemical, sorgoleone, which could negatively impact crops following sorghum as well as weeds. The research objective was to determine if sorgoleone negatively impacted germinating wheat and weeds. Lab studies were conducted to evaluate the impact of sorgoleone on growth of wheat and Italian ryegrass. Seeds of two wheat varieties and Italian ryegrass were pre-germinated and transferred to sorgoleone-treated agar in petri dishes. Sorgoleone was applied at 0, 25, 50, 100, 150, 200, and 300 μg ml^{-1}. The shoot length of treated plants were measured and compared to the shoot length of the nontreated plants 10 days after treatment. The two wheat varieties were not affected by the sorgoleone at any concentration. The Italian ryegrass shoot length was reduced at all sorgoleone concentrations. Further evaluations of five different wheat varieties from each of four type of winter wheat (soft red, soft white, hard red, and hard white) underwent the aforementioned sorgoleone tolerance screen. Sorgoleone at 300 μg ml^{-1} reduced the shoot length of the hard red and white more than the soft red and white wheat varieties. Despite the differences between the hard and soft wheat, the hard red and white wheat groups as a whole, the shoot length was not reduced more than 20% at the higher concentration which may not be detrimental to the germination wheat. Thus, the results of the experiment provide evidence that sorgoleone does negatively impact germination wheat but will suppress germinating Italian ryegrass.

WEED CONTROL RESPONSE AND CROP TOLERANCE TO FLURIDONE IN PUMPKIN. F. DAmico*, B.L. Carr, and T.E. Besançon, University of Delaware, Georgetown, DE (40)

ABSTRACT

Fluridone (Brake®) was applied immediately after seeding pumpkins at rates ranging from 42 to 672 g ai/ha. Multiple evaluations were completed for weed efficacy and pumpkin tolerance, as well as yield collection data. Rated 28 DAT, fluridone applied at ≥ 126 g ai/ha provided control of many key weeds in the trial, including common ragweed, redroot pigweed and common lambsquarters. Ivyleaf morningglory required 336 g for control while yellow nutsedge was not controlled at any rate. Weather conditions influenced crop response in all treatments including the untreated check. However, all plots treated with fluridone provided a reduced number of fruits per plot. Yield in fluridone treated plots (calculated by average fruit weight x number of fruit harvested) was reduced by >40% regardless of rate, when compared to untreated weed free check.

EVALUATION OF CHELATED FE AND OTHER SAFENERS FOR USE IN BROCCOLI. M.A. Cutulle, G.A. Caputo*, M. Farnham, B. Ward, T. Campbell, and D. Couillard, Clemson University, Charleston, SC (41)

ABSTRACT

The East Coast broccoli industry in the United States is growing and optimizing weed control strategies in the southeastern US environment is important for the expansion of commercial broccoli in the region. Based on previous studies the best time for growing broccoli in SC is in the early fall with transplants. However, nutsedge species (*Cyperus* spp.) are very active in the fall in SC and can actively compete with the broccoli transplants. Currently, no herbicides are registered in broccoli that effectively controls yellow and purple nutsedge. Bentazon is a herbicide that effectively controls nutsedge but is not registered in broccoli. The goal of this research was to find a safener that could be mixed with bentazon in order to reduce injury when applied with broccoli. A study was carried out to determine if chelated Iron and the brassinosteroid product Vitazyme effectively safened bentazon in broccoli. A greenhouse experiment was conducted as arandomized complete block, with 4 replication. The treatment design was a three level factorial, with 5 broccoli varieties, Emerald Crown, Eastern Crown, Iron Man, Belstar and BH-026 by 5 bentazon rates 0, 10, 100, 1000, 10000g ai/ha by 2 rates of Chelated Fe 0, 112 g ai/ha. A smaller experiment was conducted with vitazyme applied at 1 L/ha with bentazon to elucidate the effects of a brassinosteroid on herbicide safety. Based on the initial results it appears chelated Fe effectively safens bentazon in some scenarios.

COMPARING ANNUAL AND PERENNIAL SMALL GRAIN CROPPING SYSTEMS: AGRONOMY, ECONOMICS, AND SOIL HEALTH. E.P. Law*, S. Wayman, C. Pelzer, M.R. Ryan, and A. DiTommaso, Cornell University, Ithaca, NY (47)

ABSTRACT

Development of several perennial small grain crops has progressed to the point of commercial viability due to strong interest from multiple food industry firms, but the productivity and other impacts of perennial grain cropping systems have not been thoroughly studied in the Northeastern United States. Perennial grains may contribute to enhanced agricultural sustainability due to reduction in production inputs and increased ecosystem services such as soil and water quality protection and enhancement. Here we report preliminary results from the first two years of a field experiment comparing two perennial small grain crop cultivars, 'Kernza' intermediate wheatgrass (IWG; *Thinopyrum intermedium* (Host) Buckworth & Dewey) and 'ACE-1' perennial cereal rye (PC rye; *Secale cereale* L.), to two common annual small grain cultivars, 'Warthog' hard red winter wheat (*Triticum aestivum* L.) and 'Scala' winter malting barley (*Hordeum vulgare* L.). The two perennial crops produced less grain and more vegetative biomass than the two annuals, with grain yields of both perennials declining in the second year. An interseeded medium red clover (*Trifolium pratense* L.) companion crop improved or did not affect yield of IWG, wheat, and barley, possibly due to some combination of nitrogen fixation, soil climate moderation, and weed suppression, but outcompeted PC rye in the second year of the experiment. Partial budget analysis of the two perennial crops found that commodity prices would currently need to be about ten times greater than that of wheat to break even. Soil health and erosion control considerations in perennial grain cropping systems will also be discussed.

PRE TRANSPLANT HERBICIDE APPLICATION AND CULTIVATION TO MANAGE WEEDS IN EASTERN BROCCOLI PRODUCTION. M.A. Cutulle*, H.T. Campbell, B. Ward, M. Farnham, and D. Couillard, Clemson University, Charleston, SC (48)

ABSTRACT

The East Coast broccoli industry is growing and optimizing weed control strategies in the eastern environment is important for the expansion of commercial broccoli in the region. Research was conducted in Charleston, SC to evaluate the impact of four herbicides (pyroxasulfone, napropamide, *S*-metolachlor, and oxyflurofen) applied pre-transplant and two cultivation practices (cultivation after transplanting or no cultivation) on weed control, broccoli safety, broccoli quality characteristics and yield attributes for two broccoli hybrid cultivars, Emerald Crown and Lieutenant. The trials were conducted during the fall of 2017 and the spring of 2018. No injury was observed during the fall trial, though significant injury occurred from napropamide and pyroxysulfone application in the spring. Generally, weed competition did not significantly influence quality ratings at all. Stem diameter of broccoli was generally greater with cultivation, though likely due to enhanced aeration of broccoli roots than the added weed control. Based on the results of these studies it appears cultivation is essential for maximizing broccoli yield characteristics in coastal South Carolina while herbicide application may be less critical.

ASSESSING COVER CROP IMPACT ON WEED EMERGENCE AND VEGETATIVE GROWTH IN CABBAGE PRODUCTION. B.L. Carr*, T.E. Besançon, and J. Heckman, Rutgers University, Chatsworth, NJ (51)

ABSTRACT

Cover crops are highly efficient farming agents that promote soil health, increase water availability, help prevent erosion, and decrease the effects of pest invasions. They are also a potentially efficient method of weed suppression for organic farmers. Detailing the benefits of one cover crop versus another can aid growers as they choose the best option for their farm. Field studies were conducted in throughout the 2017 and 2018 growing seasons to test weed control potential and the impact on cabbage yield of two cover crops, sunn hemp (*Crotalaria juncea*) and sudangrass (*Sorghum x drummondii*). Trials were conducted at the Rutgers Snyder Research and Extension Farm in Pittstown, New Jersey. Cover crops were planted the year prior to crop production, were terminated through disking and tilling before cabbage planting. Weed density and biomass were evaluated at three similar calendar dates in June, July, and August of each year, and once more in September of 2018. These were done between the time the field was disked and when the cabbage crop was harvested. Cabbages were planted in late July and harvested in late October / early November. At harvest, sixteen cabbages were cut from the center two rows of each plot and weighed to evaluate yield. Sudangrass showed to have lower total weed biomass (-42%) and broadleaf weed biomass (-75%) in July 2018 in comparison to sunn hemp but had no significant effect in July 2017 and in September 2018. Sudangrass significantly decreased redroot pigweed (*Amaranthus retroflexus*) biomass collected in July 2017 and 2018 (-81 and -75%, respectively) and September 2018 (-66%) in comparison to sunn hemp. Density counts of common lambsquarters (*Chenopodium album*) were significantly lower in sudangrass than in sunn hemp plots in July 2017 (-98%) and July 2018 (-33%), but results were not significantly different by September 2018. Hairy galinsoga (*Galinsoga quadriradiata*), one of the most troublesome weed species for vegetable production in the Northeast region, showed lower biomass (-75%) and density (-80%) in sudangrass than in sunn hemp plots in September 2018. There were no significant yield differences observed in harvested cabbages between the two cover crops. While the two cover crops are both useful and comparable weed inhibitors, sunn hemp does not provide as much suppression for some weed species as sudangrass.

OVERLAPPING *S*-METOLACHLOR TREATMENTS IN LIMA BEAN. K.M. Vollmer*, M.J. VanGessel, Q.R. Johnson, and B.A. Scott, University of Delaware, Georgetown, DE (52)

ABSTRACT

More processing lima bean acreage is planted in Delaware that any other state. The onset of herbicide-resistant weeds in the area has limited herbicide options for control. Overlapping herbicides is a technique that involves sequential applications of soil-applied residual herbicides in order to overlap the herbicide's activity before the first herbicide dissipates. Separate field and greenhouse studies were conducted to evaluate weed control and lima bean response to *S*-metolachlor applied as an overlapping residual treatment.

The greenhouse study evaluated the response of six lima bean varieties to *S*-metolachlor applied to emerged lima bean plants. The study was repeated. Application rates for the first run consisted of a 1X, 2X, and 4X rate, while the second run consisted of a 2X, 4X, and 6X rate. The 1X rate of *S*-metolachlor was 801 g ha^{-1}. Lima bean injury developed within 7 days after application as leaf burn on leaves fully expanded at time of application. Less than 5% leaf burn occurred at the 1 and 2X rates. Although rate and varietal differences were observed at the 4X and 6X rates, they were not consistent across both trials. Injury symptoms were not observed on leaves that developed after the application.

The field study evaluated weed control efficacy and lima bean response to *S*-metolachlor applied PRE followed by a second application of *S*-metolachlor applied early-POST (3 weeks after planting, WAP), mid-POST (4 WAP, or late-POST [5 WAP]). Lima bean injury was greater with earlier application timings. Within a week of application, early-POST treatments injured lima bean 29%, but late-POST applications caused less than 5% injury. Injury symptoms were not observed on leaves that developed after the application. Lima bean yield was higher in treated plots, but there were no differences among treatments.

Although, weed control in all treated plots was higher compared to untreated plots, there were no differences among herbicide treatments when rated 6 WAP. Palmer amaranth control was at least 90% prior to POST treatments; however, control was 76 to 78% 6 WAP, regardless of treatment. Our results show transient injury when *S*-metolachlor is applied as a POST treatment to lima bean. However, *S*-metolachlor does not control emerged weeds. Therefore, additional research on application timing is needed to be able to provide an effective weed control strategy.

DEVELOPMENT OF A HYDROPONICS ASSAY TO SCREEN FOR *S*-METOLACHLOR TOLERANCE IN SWEET POTATO. M.A. Cutulle, H.T. Campbell*, P. Wadl, and L. Lees, Clemson University, Charleston, SC (54)

ABSTRACT

The USDA ARS Sweet Potato Breeding program and Clemson University Coastal Research and Education Center, both located in Charleston, SC, have focused on developing sweet potato varieties with more desirable traits, including tolerance to certain herbicide products. Yellow nutsedge (*Cyperus esculentus*) and Palmer amaranth (*Amaranthus palmeri*) are economically troublesome weeds in regards to sweet potato production throughout the southeastern US. *S*-metolachlor (Dual Magnum, Syngenta) is a soil-applied herbicide commonly used to suppress yellow nutsedge and Palmer amaranth in various agricultural crop settings, but concerns exist in regards to potential phytotoxic damage in sweetpotato production, as well as potential for a reduced yield in marketable sweetpotato storage roots. Therefore, an efficient method of screening sweetpotato lines to various concentrations of *S*-metolachlor is needed. Research was conducted at the US Vegetable Laboratory in Charleston, SC to evaluate the impact of *S*-metolachlor on sweet potato crop health in a hydroponic setting. The experiment was conducted as a randomized complete block design with 4 replications and 10 treatments. Two sweet potato varieties, Beauregard and Centennial, were cut into slips from 2 month old vines and transplanted into a raised bed greenhouse setting for healing. Rubber tubs contained 2 L of 1.4% Hoagland solution and each tub was connected to an air blower via pvc piping, vinyl tubing, and finally an air stone to oxygenate the nutrient solution. Slips were then harvested from raised beds after 7 days in soil and trimmed to approximately 20 cm in length before being placed in the hydroponic tub setting the same day. Two slips of the same variety were placed in each tub and secured using grafting clips. The treatments were structured as a factorial with increasing logarithmic *S*-metolachlor rates (0, .01, .1, 1, and 10 mM) x 2 varieties (Beauregard and Centennial). *S*-metolachlor was added to the tubs 1 day after the slips were placed in the tubs. Analysis did show however that *S*-metolachlor concentration significantly effects the percent phytotoxicity. Additionally, analysis also revealed a significant difference in root vigor between concentrations. This assay has to be refined but has potential to be used in the sweetpotato breeding program.

MOSS IMPACTS AND CONTROL EFFORTS IN CRANBERRY. K.M. Ghantous* and H.A. Sandler,
UMass Cranberry Station, East Wareham, MA (57)

ABSTRACT

Moss has been an increasing weed issue in cranberry cultivation. The most prevalent mosses found in Massachusetts cranberry are haircap moss (*Polytrichum commune*) and sphagnum (*Sphagnum* spp.). One bed on the UMass State Bog facility, East Wareham, MA was sampled in 2017. In addition to the *P. commune* and *Sphagnum* spp., at least three other moss species were present (*Aulacomnium palustre, Ceratodon purpureus,* and *Entodon seductrix*). A survey of a second bed at State Bog in 2018 identified at least another four moss species (*Ditrichum pallidum, Pohlia nutans, Atrichum crispum,* and *Callicladium haldanianum*). Despite the diversity of species, *P. commune* is recognized as the most common and "weedy" moss.

To assess the impact of *P. commune* on cranberry production, two beds of 'Stevens' were sampled on 27 September 2018. For each bed, fifteen 930-cm^2 squares sections of the production area were excised by using hand-held pruners to clip all plant material to the soil level. Cranberry fruit was sorted, counted, and weighted. Cranberry and moss tissue were separated, dried and weighed. Weight of sellable fruit was significantly positively correlated with cranberry plant biomass and negatively correlated with moss biomass ($p \leq 0.05$, n=30; Pearson Correlation Coefficients 0.37 and -0.40, respectively). Cranberry biomass was negatively correlated with moss biomass (Pearson Correlation Coefficient -0.43). These findings suggest direct competition between *P. commune* and cranberry.

Moss is a problem in turf and ornamentals, but not typically of concern in horticultural or agricultural crops and thus not many food-use herbicides are labeled for moss control. Mosses are non-vascular plants and do not respond to the traditional herbicides registered for use in cranberry (i.e., dichlobenil, norflurazon, napropamide, quinclorac, or mesotrione). Screening efforts have identified flumioxazin and sulfentrazone as candidates for moss control in cranberry. A set of 1-m^2 plots were established on 'Stevens' cranberry. Herbicides were delivered in the equivalent of 3,740 l/ha (400 gpa) to simulate application by chemigation. Each plot received one of nine treatments and all treatments were replicated four times. Treatments were an application of flumioxazin (Chateau, 51% a.i.) at either 280 g ha^{-1} or 420 g ha^{-1} (4 or 6 oz a^{-1}) applied early or late, or an application of sulfentrazone (Zeus XC, 39.6% a.i.) at either 560 g ha^{-1} or 840 g ha^{-1} (8 or 12 oz a^{-1}) applied early or late, or an untreated control. "Early" applications were made when cranberry plants were spring dormant (tight buds, 17 April 2018) and "late" application were made at cabbage head (buds swelling and scales beginning to loosen, 7 May 2018). Cranberry and moss were visually evaluated for injury 16 May, 25 June, and 19 September 2018. Cranberry fruit were collected on 20 September 2018 from a 930-cm^2 area in each plot, then sorted, counted and weighed. Cranberry in plots receiving applications at the later timing showed some visual symptoms of stunting on the 25 June evaluation (7 WAT); however, plants grew out of these symptoms and there were no visual cranberry differences between treatments on the final evaluation. Moss was injured by all treatments and rates. At the final evaluation, all treatments had significant moss injury compared to the control, but did not differ from each other. No treatment differed from the untreated control for number or weight of sellable fruit, even for the treatments that showed visual symptoms of injury at the June visual evaluation.

Both herbicides demonstrated good crop safety, efficacy against moss, and are registered for use in other food crops; the cranberry industry is actively pursuing registration of these herbicides.

HOW TRENDS AND EVENTS SHAPE THE ACQUISITION OF NEW MANAGEMENT TOOLS FOR CRANBERRIES. H.A. Sandler* and K.M. Ghantous, UMass Cranberry Station, East Wareham, MA (58)

ABSTRACT

Cranberries hold the unenviable position in history of being the first food crop to be the subject of a national food scare. Aminotriazole was approved for postharvest use in cranberries in 1959, the same year the Delaney Clause was enacted that stated if a substance were found to cause cancer in man or animal, then it could not be used as a food additive. At the same time, data linked aminotriazole to cancer in lab animals and to the possibility of detectable residues on cranberry fruit. The quick and intense public reaction to the news caused cranberry sales to plummet just prior to the peak distribution period (Thanksgiving market). Long-term consequences from the 1959 Cranberry Scare are felt when cranberry researchers apply for grant monies through cranberry industry partners. A waiver must be signed that commits the recipient to "not disclose or discuss" any unregistered compound as to forestall inappropriate use and/or subsequent residue detection of an unregistered product. In addition, cranberry handlers in all growing regions conduct annual pre-harvest residue testing on fruit prior to permitting delivery of the crop.

The suite of available herbicides has changed over the years. The main herbicides used during the 1940s and 1950s included kerosene, Stoddard solvent (mineral turpentine), ferrous and ferric sulfate, copper sulfate, sodium arsenate, and sodium arsenite. Starting in 1946 and ending in 1955, ammonium sulphamate and paradicholorobenzene (PDB) were available for woody weed control. Recommendation of sodium arsenite and arsenate ceased in 1955, while kerosene remained on the recommendation charts until 1990. In 2018, 18 herbicides representing 11 different modes of action were labeled for use in Massachusetts cranberries. Cranberries are grown in wetlands, which is an obstacle for the industry to obtain new products because registrants are often hesitant due to groundwater concerns.

The status of herbicide registrations can change due to regulatory changes and/or clarification of technicalities. 2,4-D (amine salt and dimethylamine salt) and a copper sulfate + lime combination were only available for one or two years in the 1950s. Triclopyr did not receive a food tolerance and its use was revoked in the same year of its approval (1983). Pronamide was available for use on MA cranberry farms through Section 18 Emergency Exemption permits from 1998 through 2007. Carcinogenicity concerns prevented additional exemptions. Subsequently, pronamide was reformulated for use on leaf lettuce and now has a reduced carcinogenicity profile; full registration for cranberry is projected to occur in 2019. Iron sulfate, used for moss control since 1937, lost its recommendation status in 2017 when regulatory authorities clarified that iron sulfate lacked technical labeling as a pesticide.

The importance of export markets has delayed the widespread adoption of new herbicides (and other pesticides) on cranberry farms, not only in MA, but throughout the U.S. Most fruits and vegetables entering foreign markets are subject to nation-established maximum residue limits (MRLs). If an MRL has not been established for a particular crop in a particular country, the usual default for the threshold is 0.01 ppm. To avoid the rejection of transported fruit to foreign markets, cranberry handlers impose restrictions on the use of certain pesticides until data-based thresholds can be established. This process can take years and significantly forestalls the use of a pesticide. For example, quinclorac was approved by the EPA for use on cranberry in 2011, but an MRL is yet to be established for the European Union (EU), a major market for exported cranberries. Since an EU-MRL is not established and handlers may not be equipped to segregate treated fruit bound for multiple export markets, quinclorac is rarely used for weed control unless the fruit is destined to stay within the U.S. (domestic market).

EFFECT OF ENVIRONMENTAL FACTORS ON VEGETATIVE GROWTH AND RHIZOME PRODUCTION OF CAROLINA REDROOT (*LACHNANTHES CAROLINIANA*). B.L. Carr and T.E. Besançon*, Rutgers University, CHATSWORTH, NJ (59)

ABSTRACT

The perennial nature of cranberry (*Vaccinium macrocarpon* L.) crop predisposes it to a diversity of weed species ranging from herbaceous weeds to woody perennial species, including Carolina redroot [*Lachnanthes caroliniana* (Lam.) Dandy] in New Jersey. Carolina redroot is a perennial herbaceous weed species belonging to the *Haemodoraceae*family. It competes for nutritional resources during the cranberry growing season, and its rhizome serve as a feeding resource for wintering waterfowl that can cause severe uprooting damages of cranberry vines when bogs are flooded. Knowledge of weed propagation mechanisms and how environmental factors may affect them remains a critical aspect of any integrated weed management program. This is especially true in cranberry production, which relies on specific cultural practices, such as sanding or flooding, that can help suppress weed emergence and development. Therefore, a greenhouse study was conducted in 2018 at the Rutgers University PE Marucci Center for Blueberry and Cranberry Research and Extension, Chatsworth, NJ, to determine the effects of various environmental factors on Carolina redroot vegetative and rhizomatic growth. The effects of planting depths (0.5, 1, 2, 4, 8, and 16 cm), soil moisture content (5, 10, 20, 30, 40, 50, and 60% of soil water-holding capacity), rhizome water content (100, 75, 50, and 30% water content), rhizome flooding duration (15, 30, 60, 120, and 180 days), and light conditions (12h day light versus complete darkness) on Carolina redroot development were evaluated by assessing daily shoot emergence as well as plant height, above- and belowground biomass 45 days after planting. Data were analyzed with NLMIXED procedure in SAS 9.4 to fit various nonlinear models. Difference in fit of the models was compared using null-model likelihood ratio and Akaike's information criteria. Based on these parameters, shoot emergence time was best fit with a parabolic model whereas cumulative percentage of emergence, vegetative growth and plant biomass were best fit with a logistic model. If timing of emergence increased with planting depth, this factor had very limited effect on Carolina redroot emergence and vegetative growth. The required planting depth for reducing emergence by 50% was between 15 and 16 cm, far beyond the 3 cm of sand applied for improving rooting of cranberry stolons. Decreased soil water content affected plant height and rate of shoot emergence with 26% and 23% soil water content, respectively, for reaching a 50% reduction of these two parameters. Reduction of rhizome water content was the most effective factor for reducing Carolina redroot emergence, plant height, and shoot biomass with required rhizome water content for reaching a 50% reduction of these parameters ranging from 53% to 57%. If rhizome submersion duration had little effect on plant height, it had a more drastic on shoot emergence and belowground biomass with 50% shoot emergence and root biomass reduction occurring when rhizomes were flooded for 143 days and 127 days, respectively. Darkness did not influence shoot emergence but significantly reduced root and shoot biomass by 75% and 53%, respectively, compared to the light conditions, and completely inhibited the development of new rhizomes.

IR-4 PROGRAM UPDATE, 2018. J.J. Baron*, D. Kunkel, R. Batts, and M.J. Arsenovic, Rutgers University, IR-4 Project, Princeton, NJ (61)

ABSTRACT

IR-4 had a successful year in 2018 with nearly 1,000 new uses posted from IR-4 research. Many of the new uses included uses on herbicides, such as clethodim, clopyralid, fomesafen, flumioxazin, sulfentrazone and others. The 2019 research program includes 16 herbicide residue studies (of a total of 65) along with supporting efficacy studies. In 2018 IR-4 initiated a new program to better service the needs of the IR-4 stakeholders, the Integrated Solutions Research Program. Integrated Solutions which is a hybrid of existing Food Use, Pest Problems without Solutions (PPWS) research and elements of the traditional Biopesticide research program. Among the six priorities selected, control of parasitic weeds in California tomato production was selected. The IR-4 biopesticide program will also be looking at weed control in organic production across fruit, vegetable and hemp production. The program continues to develop and evolve to most efficiently provide pest control solutions to US growers.

SOD REMOVAL, SPOT TREATMENT, AND SELECTIVE HERBICIDES FOR ROUGHSTALK BLUEGRASS CONTROL IN COOL-SEASON TURF. J. Brewer* and S.D. Askew, Virginia Tech, Blacksburg, VA (62)

ABSTRACT

In cool-season turfgrass systems in Virginia, roughstalk bluegrass (*Poa trivialis*) is one of the most prevalent and troublesome perennial grasses to manage on a year to year basis. Most of the herbicide options available to turf managers are inconsistent and/or ineffective. There is a need for an effective long-term chemical or mechanical option to control *Poa trivialis*. In published literature so far researchers have evaluated a range of herbicides including amicarbazone, bispyribac-sodium, glyphosate, methiozolin and sulfosulfuron with most of the research focusing on bispyribac-sodium and sulfosulfuron. Unfortunately, most of the research regarding the aforementioned herbicides have only been focused on immediate control, and only a few studies have been published to evaluate long-term *Poa trivialis* control. Researchers have observed bispyribac-sodium and sulfosulfuron controlling *Poa trivialis* greater than 80% by 8 to 12 weeks after the initial application, but very few have reported on the long-term impact of these two herbicides. Researchers in Georgia did evaluate herbicides for long-term *Poa trivialis* control. They found that even though bispyribac-sodium and sulfosulfuron controlled *Poa trivialis* effectively early in the study; *Poa trivialis* mostly recovered by the following fall or spring. Researchers have also observed glyphosate significantly reducing *Poa trivialis* cover one year after application. Interseeding desired turfgrass in combination with these nonselective and selective herbicide options have been shown to improve long-term *Poa trivialis* control compared to non-seeded programs. At Virginia Tech, research showed that methiozolin applied for two years can effectively reduce *Poa trivialis* populations by more than 80% one year after last treatment (YALT). Unfortunately, methiozolin is not yet labeled in the United States. The same study also showed that bispyribac-sodium can reduce populations between 40 and 60%, but with unacceptable injury to creeping bentgrass. In recent years at Virginia Tech, research was focused on utilizing different rates and sequential applications of primisulfuron for selective *Poa trivialis* control. One recent study evaluated primisulfuron with 18 different herbicide combinations, but no significant long-term control was observed.

This past spring at Virginia Tech a new study was initiated to focus on integrating herbicide programs with seeding perennial ryegrass at 390 kg ha^{-1}. The study was developed as a randomized complete block design with four replications. The treatments included amicarbazone at 105 g ha-1, amicarbazone plus mesotrione at 175 g ha^{-1}, glyphosate applied as a spot treatment at 10% V/V solution, glyphosate applied as a renovation treatment at 7 kg ha^{-1}, mechanical sod removal and replacement, and primisulfuron at 39 g ha^{-1}. Amicarbazone, amicarbazone plus mesotrione, and primisulfuron were applied twice at a two week interval. All herbicide treatments were applied with CO_2-powered backpack sprayer that was calibrated to spray 374 L ha^{-1} and fitted with four TTI 11006 nozzles.

By the final spring 2018 rating, amicarbazone plus mesotrione, glyphosate spot treatment, glyphosate renovation, and the sod replacement controlled *Poa trivialis* 86, 92, 100, and 100%, respectively. Primisulfuron had only controlled *Poa trivialis* 66% by this time. The evaluation in spring 2019 is the most important rating for the study to determine if any programs had effective year-long control. We hypothesize that the added turf competition from seeding and sodding will increase the long-term suppression of *Poa trivialis*. The sod replacement treatment was added to determine if a nonchemical treatment could be a viable option for homeowners.

GREEN KYLLINGA (*KYLLINGA BREVIFOLIA*) CONTROL WITH POST-EMERGENT HERBICIDES IN COOL SEASON TURF. J.M. Craft*, S.D. Askew, and J. Brewer, Virginia Tech, Blacksburg, VA (63)

ABSTRACT

Green kyllinga (*Kyllinga brevifolia*) (GK) is a rhizomatous perennial weed that is problematic in many turfgrass areas. The objective of this research was to evaluate numerous herbicides applied at different rates and timings for GK control. Three field studies were conducted in 2017 and 2018 at the Pete Dye River Course of Virginia Tech in Radford, VA on a mixture of tall fescue and creeping bentgrass mown at rough height. Herbicide treatments included sulfentrazone (Dismiss Turf), sulfentrazone + carfentrazone (Dismiss NXT), halosulfuron (Sedgehammer), sulfosulfuron (Certainty), and imazosulfuron (Celero).

At 7 days after initial treatment (DAIT), single applications of sulfentrazone and sulfentrazone + carfentrazone GK control ranged from 50 to 60%, while single applications of imazosulfuron, halosulfuron, and sulfosulfuron GK control ranged from 30 to 40%. At 14 DAIT, sulfentrazone + carfentrazone applied at 307 and 460 g ai ha, and sulfentrazone applied at 280 and 420 g ai ha controlled GK ≥ 88%, while imazosulfuron, halosulfuron and sulfosulfuron controlled GK ≤ 55%. The highest rate of sulfentrazone + carfentrazone (460 g ai ha) controlled GK at a maximum of 95%, which was similar to the highest rate of sulfentrazone. Between 28 and 35 DAIT, single applications of sulfentrazone + carfentrazone and sulfentrazone reached their maximum GK control. Two early summer applications of sulfentrazone + carfentrazone applied at a low rate (153 g ai ha) on a 2 week interval controlled GK ≥ 80% for about 42 to 48 DAIT. At 35 DAIT, single applications of imazosulfuron (420 g ai ha), sulfosulfuron (65.7 g ai ha), and halosulfuron (68 g ai ha) controlled GK ≥ 93%. Single applications of halosulfuron (68 g ai ha) and sulfosulfuron (65.7 g ai ha) reached their maximum GK control 42 and 48 DAIT. Two applications of halosulfuron (68 g ai ha) applied early summer on a 21 day interval controlled GK to commercially acceptable levels for about 70 to 77 DAIT. Imazosulfuron (420 g ai ha) applied once or twice during the early summer were the only treatments to control GK for the duration of the summer.

Sulfentrazone and sulfentrazone + carfentrazone were the fastest acting treatments but lacked the season long control. Results suggest single applications of sulfentrazone + carfentrazone and sulfentrazone GK control is similar. These data suggest applying sulfentrazone or sulfentrazone + carfentrazone at lower rates multiple times throughout the summer could increase GK control. Results indicated applying herbicides early in the summer as GK emerges is the best application timing. However, it is important to apply follow up applications to maintain season long. In conclusion, imazosulfuron was the best performing treatment regardless of application timing.

FALSE-GREEN KYLLINGA (*KYLLINGA GRACILLIMA*) CONTROL WITH AUTUMN HERBICIDE APPLICATIONS. M.T. Elmore*, D.P. Tuck, and K. Diehl, Rutgers University, New Brunswick, NJ (64)

ABSTRACT

Field experiments were conducted from 2017 to 2018 to evaluate September applications of various herbicides for false-green kyllinga (*Kyllinga gracillima*; FGK) control in Lakewood, NJ at Woodlake Country Club and in Cape May Courthouse, NJ at Stone Harbor Golf Club. The Lakewood and Cape May Courthouse locations were maintained a golf course fairway and rough, respectively. Both sites consisted of >70% false-green kyllinga cover when the experiments were initiated in 2017.

At the Cape May Courthouse site, treatments consisted of halosulfuron-methyl (70 g ha^{-1}), imazosulfuron (420 g ha^{-1}), pyrimisulfan (70 g ha^{-1}), and sulfentrazone + carfentrazone (280 + 31 g ha^{-1}) applied on 7 September 2017 and 27 June 2018 forming a four-by-two (herbicide-by-application timing) complete factorial treatment arrangement. At the Lakewood site the same herbicide treatments were applied on 14 September only forming a single-factor design. Treatments were arranged in a randomized complete block design with four replications and applied to 0.9 by 2.0 m plots. Sprayable treatments were applied using a CO_2-powered single nozzle boom with 374 L ha^{-1} of water carrier though an AI9504EVS nozzle. Imazosulfuron and halosulfuron-methyl were applied with NIS at 0.25% v/v. Pyrimisulfan was a granular formulation (EH-1617) applied using a shaker jar in the absence of dew. False-green kyllinga control was evaluated visually on a 0 (i.e., no control) to 100 (i.e., complete control) percent scale relative to a non-treated control at both locations. A grid intersect count was conducted at the Cape May Courthouse location in October 2018. Data were analyzed using the Mixed procedure in SAS and Fisher's Protected LSD (α=0.05) was used to separate means.

Imazosulfuron controlled FGK more than halosulfuron-methyl, pyrimisulfan, and sulfentrazone + carfentrazone in June 2018 at the Lakewood site. At the Cape May Courthouse location, June applications of imazosulfron and pyrimisulfan controlled FGK 98 and 56%, respectively, while September applications provided less (74 and 8%, respectively) control in October 2018 at 57 weeks after initial treatment. Halosulfuron and sulfentrazone + carfentrazone controlled FGK similarly ($\leq$ 25%) regardless of application timing in October 2018 at Cape May Courthouse. These data indicate that September applications are less effective than June applications. Future research should determine whether September applications in combination with cool-season turfgrass interseeding are more effective than June applications.

WEED CONTROL OPTIONS DURING FOUR METHODS OF ZOYSIAGRASS ESTABLISHMENT.

S. Askew* and M. Goatley, Virginia Tech, Blacksburg, VA (65)

ABSTRACT

Zoysiagrass is widely recognized for its slow establishment from sprigs, plugs, or seed. Bermudagrass, for example, requires a few months to achieve complete green cover from sprigs while zoysiagrass can require two growing seasons. Recent advancements in fraze mowing technology have led to several innovations in turfgrass agronomic management. One side effect of fraze mowing is production of large amounts of vegetative material and soil as spoils from the fraze mowing process. This debris is often discarded. Preliminary evaluations have demonstrated rapid turf establishment when bermudagrass fraze mowing spoils were spread for the purpose of establishing new turf areas. Empirical observations suggest that a 30 to 40% improvement on bermudagrass establishment time can be achieved when using fraze mowing spoils versus conventional sprigs. If this technology could be applied to zoysiagrass with similar improvement in establishment time, the technique would offer a valuable solution to the slow zoysiagrass establishment problem. During any warm-season turfgrass establishment, herbicides are needed to maximize establishment rate by excluding unwanted weeds. Our objectives were to determine zoysiagrass establishment rates from seed, plugs, sprigs, and fraze mowing spoils and to assess how uniformly-applied herbicide treatments may influence establishment rate depending on propagule type. A field study was established on May 4, 2018 and spoils, plugs, sprigs, and seed were applied at 5 m^3 ha^{-1}, 10% cover, 10% cover, and 96 kg ha^{-1}, respectively. Halosulfuron at 0.021 kg ha^{-1} and quinclorac at 0.42 kg ha^{-1} were applied on May 21, a premixture of halosulfuron + foramsulfuron + thiencarbazone was applied at 0.135 kg ha^{-1} on July 2, and quinclorac at 0.84 kg ha^{-1} was applied on August 25. Herbicides did not unacceptably injure or negatively influence establishment of zoysiagrass on any date. Zoysiagrass establishment rate exhibited the following trend from most to least rapid: seed > plugs > sprigs > spoils. Fraze mowing spoils resulted in only 10% zoysiagrass cover by November 2 while seed, plugs, and sprigs had 92, 80, and 78% cover, respectively. These data suggest that fraze mowing spoils may not be an effective method to establish zoysiagrass.

OXADIAZON PREEMERGENCE PROGRAMS FOR CONTROLLING GOOSEGRASS (*ELEUSINE INDICA*) WITH PUTATIVE RESISTANCE TO MICROTUBULE-INHIBITING HERBICIDES IN COOL-SEASON TURFGRASS. D.P. Tuck*, M.T. Elmore, S.J. McDonald, J.T. Bronson, J.J. Vargas, and G.K. Breeden, Rutgers University, New Brunswick, NJ (66)

ABSTRACT

Field experiments were conducted in New Jersey and Pennsylvania in 2018 to evaluate the efficacy of various oxadiazon programs for pre-emergence goosegrass control in cool-season turfgrass. Treatments included single oxadiazon applications (2.24, 2.80, and 3.36 kg ha^{-1}), sequential oxadiazon applications (2.24, 2.80, and 3.36 kg ha^{-1}followed by 2.24 kg ha^{-1}), single and sequential dithiopyr applications (560 g ha^{-1} and 280 followed by 560 g ha^{-1}, respectively), and single and sequential prodiamine applications (0.73 kg ha^{-1} and 0.36 kg ha^{-1}, respectively). Oxadiazon was applied as 0.67G fertilizer (0-0-50) formulation using a shaker jar. Dithiopyr (2EW formulation) and prodiamine (65WDG formulation) were applied using standard CO_2- powered small plot spray equipment.

Research was conducted at the Rutgers Horticulture Research Farm No. 2 in North Brunswick, NJ, Pinebrook Golf Course in Manalapan, NJ, and Walnut Lane Golf Course in Philadelphia, PA in 2018. The North Brunswick, NJ location was a simulated perennial ryegrass (*Lolium perenne*) fairway with known susceptibility to microtubule-inhibiting herbicides. Initial and sequential applications were made on 2 May, and 19 June, respectively, in North Brunswick. The Manalapan, NJ location was a perennial ryegrass fairway where the golf course superintendent reported poor goosegrass control from microtubule-inhibiting herbicides in previous years. Initial and sequential applications were made on 1 May, and 20 June, respectively, in Manalapan. The Philadelphia, PA location was a mixed cool-season turfgrass rough. Initial and sequential applications were made on 9 April, and 14 May, respectively, in Philadelphia. Treatments were arranged in a randomized complete block design with four replications and 0.9 by 2.1 m (NJ) or 0.8 by 1.5 m (PA) plots. Percent goosegrass control was evaluated visually relative to the non-treated control using a 0 (i.e., no control) to 100 (i.e., complete control) percent scale. Data were analyzed as a factorial with herbicide [oxadiazon (2.24, 2.80, or 3.36 kg ha^{-1}), dithiopyr, and prodiamine] and number of applications (single or sequential) as main effects. Data were subjected to ANOVA in SAS and Fisher's Protected LSD (P=0.05) was used to separate means. Pre-planned contrast statements were used to compare single versus sequential applications of oxadiazon.

At the North Brunswick location, 15 weeks after initial treatment (WAIT), all oxadiazon treatments provided similar (76 to 86%) goosegrass control. The sequential dithiopyr program and both prodiamine treatments provided 53 to 65% in North Brunswick. At the Manalapan location, all oxadiazon treatments provided ≥90% goosegrass control at 15 WAIT; sequential oxadiazon applications resulted in more control than single applications. Dithiopyr and prodiamine treatments provided poor (<10%) goosegrass control at all evaluation dates in Manalapan. At the Philadelphia location, sequential oxadiazon programs and the sequential dithiopyr program provided similar (71 to 91%) goosegrass control at 19 WAIT; sequential oxadiazon applications provided more control than single applications. Prodiamine treatments and a single dithiopyr application provided <20% control at 19 WAIT in Philadelphia. These data suggest that a goosegrass population with resistance to microtubule-inhibiting herbicides may be present at the Manalapan location, but controlled environment research is necessary for confirmation. Regardless of location, oxadiazon tended to provide more goosegrass control than dithiopyr and prodiamine.

DIMETHENAMID LEACHING IN PINE BARK VERSUS FIELD SOIL. J. Derr* and L. Robertson, Virginia Tech, Virginia Beach, VA (67)

ABSTRACT

Weed control is an important management concern in container nursery production. An important way weeds are controlled in container production is through the use of preemergence herbicides. Dimethenamid and the granular combination product pendimethalin plus dimethenamid are preemergence herbicides registered for use in this industry. The objectives of this study were to compare the weed control effectiveness of sprayed dimethenamid and the granular combination of pendimethalin plus dimethenamid in containers, to determine the leaching of these herbicides in soilless media versus field soil, and to determine the influence of irrigation volume on their leaching. Plastic 4L (1 gal) pots were filled with either 100% pine bark or with 100% field soil [Tetotum loam (fine-loamy, mixed, thermic Hapludults)], seeded with southern crabgrass [*Digitaria ciliaris* (Retz.) Koel.] or spotted spurge [*Chamaesyce maculata* (L.) Small] and then treated with either sprayed dimethenamid, granular dimethenamid plus pendimethalin, or granular pendimethalin. Soil leaching columns were constructed from 35 cm long (14 in) sections of polyvinyl chloride pipe with an internal diameter of 5 cm (2 in) and filled with either pine bark or field soil. All three herbicides provided excellent control of both weed species in field soil. In pine bark, however, dimethenamid and dimethenamid plus pendimethalin gave greater control of southern crabgrass and spotted spurge than pendimethalin. Dimethenamid leached less in pine bark than field soil, while pendimethalin was more mobile in pine bark based on a southern crabgrass bioassay. The leaching profile for the granular pendimethalin plus dimethenamid combination product was similar to dimethenamid in field soil and similar to pendimethalin in pine bark. There was no significant difference in herbicide leaching in pine bark or field soil after doubling the irrigation volume from 17.8 cm (7 in) to 35.6 cm (14 in). Compared to pendimethalin, dimethenamid leaches less in pine bark, explaining its greater effectiveness for weed control in container production. Pine bark must have a greater ability to adsorb dimethenamid than pendimethalin, as dimethenamid has greater water solubility than pendimethalin.

POSTEMERGENCE OPTIONS FOR MANAGING HORSETAIL (*EQUISETUM* SPP.). A. Senesac* and D. Barcel, Cornell Cooperative Extension, Riverhead, NY (68)

ABSTRACT

Two species of horsetail: *E. arvense* (field horsetail) and *E. hyemale (*scouringrush horsetail) can be difficult to control weeds in nurseries and sales yards. Neither species is well controlled with glyphosate. Although, these species are wide spread and native in most of North America, there occurrence as weed problems is occasional but can be severe. Their ability to spread by tubers and deep, wide-ranging rhizomes makes them formidable pests wherever they get established. Halosulfuron is registered for suppression of *E. arvense* only. There are currently no other herbicides that are commonly used in and around nurseries that are registered for managing these weeds. Two studies were conducted in 2018 to evaluate sulfentrazone in combination with halosulfuron for postemergence control of horsetail.

In New York, a container study was conducted at the Cornell University Long Island Horticultural Research and Extension Center, Riverhead, NY. Rhizomes of *E. arvense*were established in 2017 in containers and fully emerged plants treated on May 31, 2018. Sulfentrazone at 0.125, 0.25 and 0.375 lbs./a (a.i.) was applied in combination with halosulfuron at 0.0625 lbs./a (a.i). A nonionic surfactant at 0.025% was added to all treatments. Following treatment, the containers were evaluated for 147 days for percent injury. The results suggest that the horsetail foliage was blackened within a few days of sulfentrazone application. This was followed by a long period of aboveground plant disintegration. Halosulfuron when applied alone caused a very gradual and minor weakening of the plant. By the end of the evaluation period, there were still significant levels of control apparent from all three rates of sulfentrazone. This was not the case with the plants treated halosulfuron only.

In Michigan, at a commercial nursery in Kalamazoo, a mixed population of the two horsetail species was treated on July 12. A similar treatment array was applied without surfactant. The results of visual evaluations suggest that *E. arvense*, but not *E. hyem*ale was significantly injured and suppressed, especially at the higher sulfentrazone rate. *E. hyemale* was injured to some degree, but not to a level of acceptable control. In summary, while it appears that control of *E. hyemale* remains elusive with sulfentrazone, it has real promise of providing significant suppression of *E. arvense*.

EVALAUTION OF LIQUID OVER-THE-TOP HERBICIDE APPLICATION FOR FIELD GROWN HERBACEOUS PERENNIAL SOON AFTER PLANTING. H.M. Mathers*, Mathers Environmental Science Services, LLC, Gahanna, OH (69)

ABSTRACT

Herbaceous perennials (HPs) rank 3rd of 18 US major ornamental plant types sold, representing 8.6% of all national sales, and the Midwest region is tied for 1st ranking, with the Northeast, for national HP sales (Hodges et al., 2015). In herbaceous perennials one of the major production/operating costs is weed control with many HP growers/landscapers still relying upon hand-weeding. This practice constrains the industry and renders it non-sustainable in today's economy. The objectives of this trial were to find an over-the-top (OTT) liquid preemergence herbicide that would offer commercially acceptable phytotoxicity and efficacy applied alone and in combination, compared to an untreated control. Additionally, a season-long control program was evaluated with attention to site of action (SoA) rotations, using a novel winter dormant application of SureGuard WDG (flumioxazin 51%) (NuFarm Americas, Alsip, IL) (10 oz/ac) (Group 14) in Dec. 20, 2017. The SureGuard was applied across the entire area that was to be planted in May 2018 for the in-season evaluations. One newer herbicide evaluated was Tower 6EC (dimethamid-p 63.9%) (BASF Corporation, Research Triangle Park, NC) (21 oz/ac) (SoA Group 15) registered in 2011, applied with and without Dimension 2EW (dithiopyr 24%) (Dow AgroSciences LLC, Indianapolis, IN) (2 pt/ac) (Group 3). An older herbicide was also evaluated, Pennant Magnum (*S*-metolachlor) (Syngenta Crop Protection, LLC, Greensboro NC) (2 pt/ac) (Group 15) registered in 2001, applied with and without Tower 6EC. Seven species were planted as small plugs on May 7, 2018, 21 weeks after the SureGuard was applied, at Walters Gardens, Zeeland, MI, including: *Amsonia* 'Blue Ice', *Coreopsis verticillata* 'Sassy Saffron', *Sanguisorba minor* 'Little Angel', *Kniphofia thomsonii* 'Gold Rush', *Kniphofia pyromania* 'Orange Blaze', *Penstemon*'Prairie Dusk', *Penstemon* 'Midnight Masaquerade.' Shoot heights were collected at the initiation of the 2nd or in-season applications on May 16, 2018, i.e., nine days after planting. At the trial conclusion shoot height and two perpendicular measures of width were collected and put into an equation to calculate growth index values (GI). Phytotoxicity ratings were taken at four weeks at the second treatment (4WA2T) or 25 weeks after the first treatment (25 WAT), respectively and subsequently at 9 WA2T, 13 WA2T and 20 WA2T. The best herbicide for each species was based on the lowest phytotoxicity ($\leq$ 3, on a scale of 0-10, where 0 is no injury), and highest efficacy ($\geq$ 7, on a scale of 0-10, where 10 is perfect weed control). One species (*Amsonia* sp.) had no previous herbicide registrations and with the exception of Pennant on *Coreopsis verticillata*, the four herbicides tested, to our knowledge, have previously never been evaluated on the remaining five species. Commercially acceptable herbicides were found for five of the seven species. No herbicides were found for *Amsonia* 'Blue Ice' or *Penstemon* 'Midnight Masaquerade,' but reduced rates of application in further investigations may uncover an acceptable control.

EVALUATION OF CURRENT HERBICIDES FOR PHYTOTOXICITY AND EFFICAY OF FIVE HERBACEOUS PERENNIAL SPECIES IN CONTAINERS. H.M. Mathers*, Mathers Environmental Science Services, LLC, Gahanna, OH (70)

ABSTRACT

The key to the obtaining higher returns in high demand herbaceous perennial (HP) crops is primarily dependent on lowering weed control costs. In addition to lack of herbicide registrations for HP crops, lack of access to workers, and the increased expense of such labor, if available, has diminished the growth potential of this sector. Therefore, the development of new and improved herbicides and label expansion work with common and high value HP crops is required. Unlike a common woody nursery crop like *Berberis* sp. which has 17 registered herbicides, a common HP crop like *Lobelia* sp. has five (Stamps et al., 2012). At Ray Wiegand's Nursery, Lenox, MI five species were evaluated with eight treatments applied 03/30/2018. Each phytotoxicity and efficacy mean represents eight replications of one-gallon containerized herbaceous plants. Phytotoxicity and efficacy ratings were taken at 6WAT on May 11, 2018, 11 WAT on June 13, 2018 and 16 WAT, July 19, 2018. The objectives of this study were to evaluate efficacy and phytotoxicity of three newer granular herbicides, compared to liquid applications of Tower 6EC (dimethamid-p 63.9%) (BASF Corporation, Research Triangle Park, NC) (26 and 52 oz/ac) (Group 15) registered in 2011, applied with and without Dimension 2EW (dithiopyr 24%) (Dow AgroSciences LLC, Indianapolis, IN) (2 pt/ac) (Group 3) and an untreated control. The three granular formulations evaluated included: Fortress (isoxaben 0.50% + dithiopyr 0.25%) (OHP, Inc., Mainland, PA) (150 and 300 lb/ac) (Group 21 + 3) registered in 2018; Biathlon (oxyfluorfen 2% + prodiamine 0.75%) (OHP, Inc., Mainland, PA) (100 lb/ac) (Group 14 + 3), and Marengo G (indaziflam 7.4%) (Bayer Crop Science Inc, Research Triangle Park, NC) (200 lb/ac) (Group 29) both which garnered EPA registrations in 2013. The five species evaluated included, *Rudbeckia fulgida* 'Little Goldstar', *Penstemon schmidel* 'Red Riding Hood', *Panicum virgatum* 'Shenandoah', *Iris sibirica* 'Sparkling Rose', and *Asclepias incarnata.* The best herbicide for each species was based on the lowest phytotoxicity ($\leq$ 3, on a scale of 0-10, where 0 is no injury), and highest efficacy ($\geq$ 7, on a scale of 0-10, where 10 is perfect weed control). One species (*Panicum virgatum* 'Shenandoah'.) had no previous herbicide registrations and the remaining four species, to the best of our knowledge, are previously untested with the five products evaluated in this trial. Commercially acceptable herbicides were found for five of the five species. For *Iris sibirica*'Sparkling Rose', and *Asclepias incarnata* multiple herbicides were acceptable with 4 for *Iris* and 5 for *Asclepias*. *Rudbeckia fulgida* 'Little Goldstar' was the only species with only one acceptable control identified and the remaining two species had two controls each. Fortress and Tower (26 oz) were commercially acceptable for four of five species, but not the same four species. Fortress is currently the broadest HP label in the marketplace with 131 HP crops and groundcovers. Fortress also shows exceptional potential for further label expansion.

CONTROLLING GLYPHOSATE-TOLERANT *CONYZA* IN NC FRASER FIR PRODUCTION. J. Neal*, C. Harlow, and J. Owen, North Carolina State University, Raleigh, NC (71)

ABSTRACT

In NC Fraser fir production fields, a living ground cover dominated by white clover is maintained to prevent erosion. Control options are needed that will remove *Conyza* and other escaped weeds without reducing that clover cover. Experiments were conducted at two Fraser fir Christmas tree farms and in containers to evaluate the safety and efficacy of herbicides for selective control of glyphosate-tolerant *Conyza canadensis*. Herbicides evaluated in this study included cloransulam-methyl (FirstRate) at 0.3, 0.6 and 1.2 oz/A + 0.25% v/v NIS; and thifensulfuron-methyl (Harmony) at 0.25, 0.5 and 1 oz/A + 0.25% v/v NIS; 2,4-D dimethylamine at 33.7 oz/A; saflufenacil (Detail) at 0.5 and 1 oz/A + 1% v/v MSO; imazapic (Plateau) at 4 and 6 oz/A + 0.25% v/v NIS; topramazone (Frequency) at 4 and 8 oz/A + 1% v/v MSO; and glyphosate (Razor Pro) at 4 or 8 oz oz/A. Herbicides were applied as a directed spray to the lower 12 to 15 inches of the trees with a CO_2 pressurized sprayer equipped with two TeeJet 11002 TTI drift-reducing nozzles and calibrated to deliver 15 GPA. Treatments were applied in a randomized split-block design with 4 replicates. Main plots were herbicide treatments; sub-plots were 3 application timings: at or pre-bud break; when shoots were round to early 'V' stage; or when branches were elongated, with "flat" needles. There were 5 trees in each sub-plot. Plant injury was visually evaluated on a 0 to 10 scale where 0 = no visible injury and 10 = complete necrosis of treated branches. Plant injury evaluations focused on the treated branches, not the entire plant. *Conyza* control was evaluated at one field site and in a container experiment in Raleigh, NC.

Although there was some slight chlorosis on a few plants from cloransulam-methyl or glyphosate applications, the level of injury was inconsequential. No injury was observed from thifensulfuron-methyl or topramazone. Saflufenacil or 2,4-D amine applied at bud break did not injure trees, but applications after bud break caused significant injury. Imazapic applied pre- or post-bud break resulted in systemic chlorosis of the treated shoots and terminals. Of the treatments tested, only topramazone killed the clover ground cover. However, it should be noted that clover populations in these fields were not uniform enough to fully assess herbicide impact on ground cover suppression and recovery. Neither glyphosate nor thifensulfuron-methyl nor imazapic controlled horseweed. All cloransulam-methyl + NIS, topramazone + MSO, and 2,4-D amine applications controlled horseweed. In the field test, saflufenacil applied pre-bud break at 1 oz/A + MSO controlled horseweed 100%, but it was less effective at later growth stages and in the container experiment. Use of the recommended spray adjuvants improved *Conyza* control with cloransulam-methyl and topramazone, compared to those herbicides tank mixed with glyphosate without the addition of spray adjuvants. Omitting MSO from the saflufenacil + glyphosate tank mix did not reduce efficacy on *Conyza*. These data support continued research on the registration of cloransulam-methyl and thifensulfuron-methyl and continued testing of pre-bud break applications of saflufenacil and 2,4-D to provide growers with resistance management rotation options in Fraser fir Christmas tree production.. In grower practice, these herbicides would likely be tank mixed with glyphosate for suppression of grasses and other vegetation.

SHORT-TERM EFFECTS OF BASAL STEM GLYPHOSATE APPLICATIONS TO NURSERY TREES. A. Witcher*, Tennessee State University, McMinnville, TN (72)

ABSTRACT

Glyphosate is the most common post-emerge herbicide for controlling weeds in field-grown nursery crops. Glyphosate is sprayed as a band application to terminate weeds within the row, but the herbicide can contact tree stems when non-shielded spray tips are used. Glyphosate can be absorbed by tree bark and translocated within the plant, but the potential damage caused by these applications has not been extensively studied. The objective of this research was to evaluate basal glyphosate applications on growth of 'Summer Red' red maple (*Acer rubrum* L. 'Summer Red') over two growing seasons. Two experiments were conducted in summer 2018 at the Tennessee State University Nursery Research Center in McMinnville, TN using 1 gal container-grown 'Summer Red' red maple plants. In Experiment 1, plants received basal (15 – 20 cm portion of lower stem) glyphosate treatments every 4 or 8 weeks (4 or 2 total applications, respectively) using an application volume of 2, 20, or 40 gallons per acre (1 lb ai/A). In Experiment 2, plants received basal glyphosate treatments every 4 or 8 weeks (4 or 2 total applications, respectively) using an application rate of 1 or 2 lb ai/A glyphosate. In both experiments, an ultra-low volume sprayer (Herbiflex 4) was also included as a treatment for comparison with conventional sprayer applications. Shoot height was recorded at the beginning and end of the experiments, while shoot biomass was measured 4 months after the initial treatment.

No visual symptoms of herbicide damage were observed in either experiment. In Experiment 1, change in plant height and shoot biomass were not significantly affected by application volume or number of applications. In Experiment 2, no differences in change in height or shoot biomass were observed for glyphosate rate or number of applications. Although glyphosate can be absorbed through bark and translocated throughout the plant, visual and residual effects may take longer to materialize compared with leaf and exposed root absorption. In our experiments, half of the plants (6 replications) were harvested for shoot biomass while the remaining plants will be transplanted to 3 gal containers and receive the same glyphosate treatments in spring 2019. In the short term, basal glyphosate applications do not seem to damage young red maple trees but second year growth will be measured to quantify residual effects.

EXTENSION EFFORTS OF GROW (GETTING RID OF WEEDS): A NATIONAL TEAM RESEARCHING AND DRIVING ADOPTION OF INTEGRATED WEED MANAGEMENT. M.L. Flessner*, M.J. VanGessel, K.B. Pittman, C. Rubione, S. Mirsky, M. Bagavathiannan, and L.M. Lazaro, Virginia Tech, Blacksburg, VA (73)

ABSTRACT

GROW is a USDA Agricultural Research Service (ARS) area-wide funded integrated weed management (IWM) team of researchers and extension specialists. The project "An integrated pest management approach to addressing the multiple herbicide-resistant weed epidemic" includes traditional research, social and economic analyses, and outreach and technology transfer (i.e. extension) components.

Extension efforts are multifaceted and aimed at increasing knowledge and adoption of IWM practices. To focus subsequent efforts, a national IWM survey was conducted to gauge current understanding and adoption of IWM across the US. The survey data and literature indicate that many farmers are routinely using some IWM practices, many farmers are unclear about multiple, effective herbicides sites of action, and many farmers are unsure about herbicide use combined with alternative management tactics. Literature further indicates that best practices to accelerate adoption should (1) focus on a tactic that is easy to adopt, (2) consist of compelling data, and (3) be delivered to a targeted audience to leverage farmer preference for peer-to-peer learning.

Additional extension efforts include online content in the form of a website (integratedweedmanagement.org) and the social media platforms: Twitter (@GetRidofWeeds), Facebook (facebook.com/GetRidofWeeds/), and YouTube (GROW IWM channel). The website provides a centralized source of reliable IWM information that includes existing/linked content as well as original content. The website also hosts a question/answer forum on IWM, allows users to subscribe to articles and updates, and catalog herbicide-resistant weeds in each state. Social media platforms have a three-part objective: drive users to the website, encourage peer-to-peer interactions, and find and connect to early adopters.

Lastly, we are developing two decision support tools. Effective site of action (ESOA) will be an app and web-based tool that reports the number of ESOA, risk categories for herbicide resistance, and herbicide product alternatives to increase ESOA, after the user inputs their crop, dominant weed(s), and herbicide(s) or herbicide program. Palmer Amaranth Management (PAM; https://agribusiness.uark.edu/decision-support-software.php#PAM) is a decision support tool to evaluate management practices on Palmer amaranth (*Amaranthus palmeri*) seedbank dynamics and economic returns. PAM will be expanded to: include research and educational modules for crop rotation, tillage, and other practices; increase the geographic region of inference; and link to other open-source decision support platforms.

ASSESSING WEED EMERGENCE MODELS IN THE NORTHEAST. C. Marschner*, M.J. VanGessel, A. DeGaetano, and A. DiTommaso, Cornell University, Ithaca, NY (74)

ABSTRACT

Extensive research has been conducted to create models for weed emergence in a range of crops, regions, and conditions. Decision support tools for weed management have been created in several agricultural regions, but a tool for the Northeast is lacking. The Weed Ecology and Management Lab at Cornell University has collected two years of data to test central Atlantic and Midwestern hydrothermal time equations and soil moisture cutoffs for central New York. Preliminary data indicate that different models are suited to particular species and weather conditions, and that additional work to fit the model to a range of soil types is needed. While model fit was strong for some species in one year or the other, no model predicted emergence of any species at both years and sites. In drought periods, the current modeled soil moisture cutoffs for growing degree day accumulation reliably overestimated weed emergence in gravelly soils. In plots where soil was disturbed to mimic cultivation, emergence was slower in dry periods.

In 2018, a multistate partnership with participants ranging from northern New England to Virginia received funding to extend this research across the Northeast. Our goal is to strengthen model fit across the Northeast, and develop a decision support tool to be hosted on the Northeast Environment and Weather Applications (NEWA) website.

EFFECT OF COVER CROPS ON POTATO YIELD AND QUALITY. J.M. Jemison*, University of Maine, Orono, ME (75)

ABSTRACT

Soil management through cover cropping or undersowing can benefit potato yields in the following year by capturing and holding residual nitrogen (N), reducing erosion and N leaching losses, but can also possibly reduce yield through competition for water and nutrients if not effectively killed. Many potato growers want a clean field to prepare the potato seed bed, but this can lead to soil and nutrient loss. In 2017, an experiment was initiated to assess the effect of cover crops on potato yield and quality. Barley was planted in May 2017 with and without undersown annual rye grass (AGR). After harvest, AGR was disked in late August or early November or sprayed with glyphosate in early November. In plots without undersown AGR, oats or winter rye was sown in September to provide winter cover. All plots were disked in May 2018 and planted to Kennebec potato. Two of the AGR treatments did not winter kill over the 2017/2018 winter, and those treatments had to be disked in the spring. Even then some AGR survived and competed with potatoes for water and nutrients. The oats winter killed and did not compete. Winter rye was effectively killed by two passes with the disk. Effect of AGR and cover crops on plant growth and development and marketable and total yield will be presented.

WHAT'S NEW. S.A. Mathew*, D. Porter, D. Black, S. Cully, and G. Letendre, Syngenta, Germantown, MD (76)

ABSTRACT

Tavium® Plus VaporGrip® Technology herbicide is a premix of *S*-metolachlor, dicamba, and VaporGrip Technology® that is under regulatory review for registration by the U.S. Environmental Protection Agency (EPA). Upon EPA approval, it will control key broadleaf weeds and grasses in Roundup Ready 2 Xtend® Soybeans or Bollgard II® XtendFlex® Cotton.

Gramoxone Magnum™ is a premix of paraquat and *S*-metolahclor also under regulatory review awaiting EPA approval. Upon registration, Gramoxone Magnum™ will provide fast, reliable control of emerged weeds and residual control of broadleaf weeds and grasses. It is expected to be registered for use in corn, cotton, legume vegetables, sorghum, soybean, sunflower, and post-harvest weed management.

Axial® Bold is a new cereals herbicide from Syngenta that contains the active ingredients pinoxaden and fenoxaprop. This combination delivers improved consistency and broad-spectrum post-emergence control of troublesome grasses such as wild oat, foxtail species, Italian ryegrass, and barnyardgrass. Enhanced flexibility makes it a convenient tank mix partner with broadleaf herbicides like Talinor® for broad-spectrum control of both grasses and broadleaf weeds.

Tavium® Plus VaporGrip® Technology and Gramoxone Magnum™ are not yet registered for sale or use in the United States and are not being offered for sale. Upon registration, Tavium and Gramoxone Magnum will be Restricted Use Pesticides. Gramoxone SL and Gramoxone SL 2.0 are Restricted Use Pesticides.

PALMER AMARANTH CONTROL IN ALFALFA/GRASS MIX: ANY HERBICIDE OPTIONS? D. Lingenfelter*, Penn State, University Park, PA (77)

ABSTRACT

Palmer amaranth (*Amaranthus palmeri*) was recently introduced into Pennsylvania agricultural systems. It was initially documented on several farms in 2013 and continues to spread throughout the state. Most infestations are primarily observed in soybean, corn, alfalfa, and around barnyards, however it has invaded other crops and areas as well. Practical management options (i.e., herbicides, cover crops, crop rotations, tillage, equipment sanitation, etc.) are available in corn and soybean production systems and are commonly recommended to farmers. However, there are limited management suggestions (namely herbicides) in alfalfa, especially alfalfa/grass mixtures. About 60% or more of the alfalfa acres in the state are mixed with a forage grass species; in some cases, to thicken the stand for improved competition with weeds. In pure-stand alfalfa, paraquat combined with pendimethalin or flumioxazin has been recommended soon after cutting to provide foliar control of Palmer amaranth stump regrowth and residual control of new seedlings; however, these combinations are not permitted currently in an alfalfa/grass mixture. Furthermore, glyphosate-tolerant alfalfa varieties are being utilized more in the state allowing over-the-top application of glyphosate to control problem weeds, unfortunately most of the Palmer amaranth in the region is glyphosate-resistant, thus making this option impractical. Because of these issues, initial investigation of potential concept herbicide treatments for Palmer amaranth control after first or second cutting in alfalfa/grass have been conducted. A field study was established in 2018 at Rock Springs, Pennsylvania to examine paraquat (0.25 lb ai/A), glufosinate (0.53 lb), and fomesafen (0.25 lb) alone and in combination with pendimethalin (1.9 lb) and flumioxazin (0.096 lb) applied soon after the first cutting in alfalfa (*Medicago sativa*)/orchardgrass (*Dactylis glomerate*) to determine their impact on crop injury and yield. Treatments were applied on May 31 to 2-inch alfalfa regrowth and 3-inch orchardgrass regrowth. Necessary adjuvants were included in the treatments. Visual crop injury ratings were collected periodically, and forage yield was taken on July 9. (No Palmer amaranth population was at this site; thus, no control data was collected.) Small-plot studies were arranged in a randomized complete block design with two replications. Data from the last evaluation reveal that in general, treatments containing fomesafen caused the least injury to both crop species (alfalfa: 3-5%; orchardgrass: 0-3%); those with paraquat caused intermediate injury to orchardgrass (alfalfa: 3-10%; orchardgrass: 18-23%); and those with glufosinate caused the most crop injury especially to orchardgrass (alfalfa:5-13%; orchardgrass: 45-50%). Crop yields from each treatment followed a similar trend – control: 1.6 tons/A; fomesafen: 1.36-1.54 tons; paraquat: 1.08-1.28 tons; and glufosinate: 0.7-0.76 tons. Preliminary data suggests that fomesafen and paraquat might have some utility for this situation, but not glufosinate. Further studies need to be conducted on these herbicides (and possibly others) to collect additional data on weed control efficacy and crop injury potential especially for various forages grass species (e.g., orchardgrass, timothy, bluegrass, ryegrass, bromegrass, etc.) that might be used in alfalfa/grass mixes. If subsequent data shows positive results, then recommendations to companies to include this utility on specific herbicide labels should occur.

EVALUATING MANAGEMENT TRADEOFFS IN FULL- AND NO-TILLAGE ORGANIC SOYBEAN PRODUCTION. J.M. Wallace*, R. Champagne, and W.S. Curran, Pennsylvania State University, University Park, PA (78)

ABSTRACT

Organic grain producers are interested in reducing the frequency and intensity of tillage to improve soil conservation and decrease labor and fuel costs. A cropping system experiment was conducted between 2015 and 2017 to examine alternative strategies for reducing tillage frequency and intensity within a corn – soybean – spelt rotation in the northern Mid-Atlantic. In the soybean phase, full-tillage soybean production, which was preceded by use of a cover crop mixture interseeded into corn grain, was compared to no-till soybean production, which was preceded by fall sown cereal rye after corn silage. Full-tillage soybean were planted into a prepared seedbed on 76-cm row spacing at 444,000 plt ha^{-1} and no-tillage soybean were planted into roller-crimped cereal rye on 76-cm row spacing at 555,000 plt ha^{-1}. False seed bedding, blind cultivation and inter-row cultivation were employed adaptively to manage weeds in the full tillage system and high-residue cultivation was employed adaptively to manage weeds in the no-tillage system. Mean late-season weed abundance was not statistically different between soybean systems but showed high inter-annual variability. Mean soybean yields were approximately 0.28 Mg ha^{-1}higher in the full-tillage soybean system compared to the no-till soybean system across study years. Path analysis results showed that soybean establishment was the greatest yield-limiting factor in both soybean systems. Evaluation of soil disturbance intensity from field operations using STIR index values showed that the no-till soybean system resulted in 50 %less soil disturbance compared to the full-tillage production system across study years. Finally, enterprise budget comparisons between soybean production systems showed that no-till soybean production resulted in lower input costs than full-tillage production methods, but no-till soybean was $114 ha^{-1}less profitable averaged over study years compared to full-tillage production due to lower revenues.

PALMER AMARANTH CONTROL IN GRAIN SORGHUM OR SOYBEANS: DOES IT MATTER?

M.J. VanGessel*, Q.R. Johnson, and B.A. Scott, University of Delaware, Georgetown, DE (79)

ABSTRACT

Weeds can severely reduce grain yields if not adequately controlled. Palmer amaranth is an aggressive summer annual weed capable of causing significant yield loss in all summer planted crops. It has the ability to germinate throughout the summer, but a dense crop canopy can suppress Palmer amaranth growth. A better understanding of the role crop rotations have in weed management is needed to provide advice on managing this species. Limited research has examined sorghum as an alternative crop for drought-prone fields infested with Palmer amaranth. This research was designed to examine comparable sorghum and soybean herbicide programs for successful Palmer amaranth control.

The study was initiated in 2018 as a strip trial with sorghum or soybeans as the whole plot and herbicide programs as the subplot. The six herbicide programs were 1.) *S*-metolachlor applied PRE; 2.) *S*-metolachlor plus triazine applied PRE (atrazine [sorghum] or metribuzin [soybeans]); 3.) three-way PRE herbicide program (treatment 2 plus mesotrione [sorghum] or flumioxazin [soybean]); 4.) *S*-metolachlor applied PRE followed by dicamba POST; 5.) PRE followed by POST (treatment 2 followed by pyrasulfotole+bromoxynil+atrazine [sorghum] or fomesafen [soybeans]); and 6.) nontreated check.

Crop stunting was minimal for all soybean treatments; all sorghum treatments containing atrazine resulted in 11% stunting at 12 days after planting (DAP). Main effects were significant for Palmer amaranth control at all rating dates. Sorghum always had a higher level of Palmer amaranth control compared to soybeans, ranging 7 to 10% higher. At 36 DAP, the three-way PRE program and both POST programs provided the highest level of Palmer amaranth control. At harvest, only the two programs with a POST application provided the highest level of control. Yield was similar for all treatments. When yield as calculated as a percent of the untreated check, there was no difference between sorghum and soybeans or for any herbicide program.

Based on this first year trial, sorghum may provide better Palmer amaranth options than soybeans, presumably due to its increased competitive ability. Additional research is needed to determine the consistency of these results; and to examine the nature of improved Palmer amaranth control with sorghum.

EVALUATION OF POPULATION AND GROWTH RATE OF PALMER AMARANTH FOLLOWING VARIOUS PRE-EMERGENT RESIDUAL HERBICIDES IN SOYBEANS. B. Beale*, University of Maryland, Leonardtown, MD (80)

ABSTRACT

Palmer amaranth (*Amaranthus palmeri*) continues to present a major challenge to crop producers in Maryland, particularly in soybean systems. Field trials were conducted for three years at two locations to evaluate preemergence and delayed preemergence herbicide treatments for control of Palmer amaranth. Trials were conducted at two locations in Southern Maryland, one in Charles County near the town of Newburg and the second in St. Mary's County near the town of Mechanicsville. Each site had a heavy infestation of glyphosate and ALS resistant Palmer amaranth the preceding crop year. Treatments consisted of (metribuzin + chlorimuron (Canopy®); flumioxazin + *S*-metachlor (Valor SX® + Dual Magnum®); flumioxazin (Valor SX®); sulfentrazone + chlorimuron (Authority XL®); sulfentrazone + *S*-metolachlor (BroadAxe XC®); *S*-metolachlor + metribuzin (Boundary®); metribuzin (Dimetric®); pendimethalin (Prowl H2O®); flumioxazin + pyroxasulfone (Fierce®); flumioxazin + pyroxasulfone + chlorimuron (Fierce XLT®); sulfentrazone + *S*-metolachlor (BroadAxe XC®) fb *S*-metolachlor (Dual Magnum®) 21 days after treatment (dat); flumioxazin + pyroxasulfone (Fierce®) fb *S*-metolachlor (Dual Magnum®) 21 dat; *S*-metolachlor + metribuzin (Boundary®) fb *S*-metolachlor (Dual Magnum®) 21 dat; and pendimethalin (Prowl H2O®) + metribuzin (Dimetric®) *S*-metolachlor (Dual Magnum®) 21 dat. Preemergent treatments were applied within three days after planting. Delayed preemergence treatments were applied 21 days after initial treatment with the preemergence residual. A complete block randomized design with four replications was utilized at each location. Before planting, all existing vegetation was terminated with one to three applications of paraquat and plots monitored to ensure a clean field seed bed prior to planting. Counts of emerged Palmer amaranth plants were taken from the middle 4 rows of each plot, which represented an actual area of 13.9 m^2 (1.524 m by 9.14 m). The number of emerged Palmer amaranth plants was recorded from individual plots starting 10 days after treatment and every 7 days thereafter. The average height of the five tallest emerged Palmer amaranth plants in each plot was also recorded. Data was analyzed in JMP 11 Pro using ANOVA and means separated for significant differences at the 0.05 confidence level.

Results indicate significantly better control of palmer amaranth with the use of any residual product when compared to a non-treated control. Residual products varied in the length of control provided. Products with the active ingredient flumioxazin or products with the active ingredient sulfentrazone provided the most consistent control. Treatments also resulted in significant differences in weed height over the length of the season. Palmer amaranth is most effectively controlled with postemergent herbicides when plants are less than 10 cm tall. Results indicate a post-emergent herbicide treatment needs to be applied within 24 to 28 days of the initial effective preemergence residual treatment to achieve reliable control. If no residual herbicide is used, the post-emergent treatments need to be applied within 14 days, and season long control is very unlikely.

TAVIUM™ PLUS VAPORGRIP® TECHNOLOGY- A TOOL FOR WEED MANAGEMENT IN CONVENTIONAL AND NO-TILL DICAMBA TOLERANT SOYBEANS. L. Smith*, A. Franssen, B. Miller, T. Beckett, and D. Porter, Syngenta, King Ferry, NY (81)

ABSTRACT

Tavium® Plus VaporGrip® Technology is a new herbicide premix developed by Syngenta Crop Protection for use in dicamba tolerant soybean and pending registration with the EPA. It will contain three key components: dicamba, a Group 4 herbicide, *S*-metolachlor, a Group 15 herbicide, and VaporGrip Technology which decreases the volatility of dicamba and reduces the chance for off-site movement. Upon registration, Tavium Plus VaporGrip Technology will provide postemergence control of over 50 broadleaf weeds as well as extended residual control of key broadleaf species such as waterhemp and Palmer amaranth as well as troublesome grasses. Tavium Plus VaporGrip Technology will offer flexibility in application timing by allowing one application from preplant burndown through preemergence and one application postemergence in dicamba tolerant soybean. By employing two modes of action, Tavium Plus VaporGrip Technology will be an effective resistance management tool which will fit well into either conventional or no-till systems by delivering postemergence control and enabling overlapping residual activity.

HORSEWEED (*CONYZA CANADENSIS*) CONTROL WITH PRE-PLANT HERBICIDES AFTER MECHANICAL INJURY FROM SMALL GRAIN HARVEST. M.L. Flessner* and K.B. Pittman, Virginia Tech, Blacksburg, VA (82)

ABSTRACT

Horseweed is a troublesome weed and can be difficult control due to multiple herbicide resistance. Response to herbicides applied at various horseweed growth stages is well characterized, but little information is available on horseweed that has been mechanically cut during small grains harvest and subsequently treated. The objectives of this research were (1) to evaluate herbicidal control of decapitated horseweed and (2) to compare herbicide efficacy when applied to rosette, bolt, and decapitated horseweed.

In 2015 and 2016, field research evaluated herbicidal control of glyphosate-susceptible horseweed that was cut to 46 cm tall to simulate a combine harvester 1 to 3 d prior to treatment. A randomized complete block design with a minimum of 3 replicates per treatment was used, with plot sizes of 2.4 by 6 m. Treatments included glyphosate at 1.26 kg ha^{-1}, 2,4-D at 1.12 kg ha^{-1}, dicamba at 0.56 kg ha^{-1}, glyphosate + 2,4-D, glyphosate + dicamba, paraquat at 0.84 kg ha^{-1}, paraquat + 2,4-D, paraquat + dicamba, glufosinate at 0.595 kg ha^{-1}, and a nontreated check. Treatments were applied using a CO_2-pressurized hand-held spray boom equipped with four TeeJet VS8002XR nozzles with 46 cm spacing calibrated to deliver total spray volume of 140 L ha^{-1} at 207 kPa. Visible control ratings were taken relative to the nontreated check on a 0 (no control) to 100 (complete plant necrosis) scale 2, 4, and 6 weeks after treatment (WAT). ANOVA was performed, and effects were considered significant when $p < 0.05$. Subsequently, means were separated using Fisher's protected LSD (α=0.05). Additionally, greenhouse experiments in 2016 to 2018 evaluated glyphosate-resistant horseweed with two factors: horseweed status at application (rosette and bolt growth stages and decapitated) and herbicide treatment. Treatments, application, data, and analyses were identical to the field study, with the addition of above ground biomass collected 6 WAT.

Paraquat, alone or in combination with 2,4-D or dicamba, resulted in 80 to 100% control of horseweed that was decapitated prior to treatment across field and greenhouse experiments. Other treatments resulted in less decapitated horseweed control in at least one experimental run. Dicamba and 2,4-D controlled rosette stage glyphosate-resistant horseweed >90%, consistent with previous research. Across herbicides, the rosette growth stage was most effectively controlled, followed by the bolt growth stage, which was similar or more effectively controlled than decapitated horseweed. This research agrees with previous reports and recommends treating horseweed in the rosette growth stage. Future research is needed to evaluate these treatments *in situ*, rather than simulating small grains harvest.

GRAMOXONE MAGNUM: A NEW OPTION FOR BURNDOWN AND RESIDUAL WEED CONTROL. E. Hitchner*, D. Bowers, R. Lins, A. Moses, and M. Saini, Syngenta, 08098, NJ (83)

ABSTRACT

Gramoxone Magnum herbicide is a new product for burndown and residual control of grass and broadleaf weeds in corn, legume vegetables, sorghum, soybeans, and sunflower. Gramoxone Magnum is a combination of paraquat (Group 22) and *S*-metolachlor (Group 15). Upon EPA approval, it will provide two alternative sites of action to glyphosate (Group 9) and has tank mix flexibility for multiple cropping systems.

HARNESS® MAX HERBICIDE: A FLEXIBLE PRODUCT FOR WEED MANAGEMENT IN CORN.

R.A. Recker*, D.J. Mayonado, R.F. Montgomery, C.M. Mayo, and G.G. Camargo, Bayer Crop Sciences, St. Louis, MO (84)

ABSTRACT

Harness® MAX Herbicide is a flexible corn product that is a pre-formulated mixture containing mesotrione and acetochlor with the safener furilazole. Harness® MAX Herbicide offers the excellent residual benefits of acetochlor with the added post-emergence and residual activity of mesotrione for a broadened range of control against tough to control weeds in corn such as amaranths (A*maranthus sp.*), common lambsquarters (*Chenopodium album*), morningglories (*Ipomoea sp.*) and foxtails (*Setaria sp.*). Field studies were conducted in 2018 to evaluate residual weed control and crop response of Harness® MAX Herbicide compared to competitive herbicides. In one study, applications were made pre-emergence and in a second study applications were made post-emergence, targeting 2-4 inch weeds or 11 inch corn, whichever occurred first. Overall, results from these studies indicate that Harness® MAX Herbicide can provide excellent residual weed control and that this convenient pre-mix herbicide can be a valuable weed management tool for corn growers.

BANDED AND IN-FURROW ACTIVATED CHARCOAL IMPROVES WILDFLOWER TOLERANCE TO PREEMERGENT HERBICIDES. S. Askew* and M. Shock, Virginia Tech, Blacksburg, VA (85)

ABSTRACT

Pollinator-serving plants have become a popular landscape addition in recent years. Most plants used for this purpose are slow-establishing perennials. Successful pollinator gardens often require intensive hand removal of weeds during establishment or several years of successional displacement of unwanted weeds. Traditionally, leaders in the pollinator plant industry discourage the use of synthetic herbicides. Thus, limited research has evaluated herbicides as a tool to aid establishment of pollinator-serving plants. As desire for pollinator services moves more mainstream, professionals in the landscape industry need methods for rapid plant establishment and pest control. Preemergence herbicides would be ideal for reducing weed pressure during the slow pollinator plant establishment. Species diversity and variable seedling vigor of pollinator plants, however, reduces application of preemergence herbicides to aid plant establishment. Activated charcoal applied in narrow bands over planted rows can impart safety to desired plants and allow preemergence control of weeds in row middles. To test this technique for a variety of pollinator-serving plants, greenhouse studies were conducted in Blacksburg, VA. Our objective was to determine the influence of four activated-charcoal treatments on four pollinator-serving plant's response to diuron, indaziflam, and pendimethalin. Activated charcoal powder was mixed with 0.1% w/w xanthan gum and suspended in water at 13.9 kl solution per kg charcoal powder. The charcoal solution was applied to native silt-loam soil mixed 2:1 with Profile Greens-grade ceramic in 1 dm^2 pots. Charcoal treatments included: none, 336 kg/ha over the soil surface (OT), 672 kg/ha OT, and 672 kg/ha half OT and half in furrow. After applying charcoal, the following herbicides treatments were applied: none, diuron at 4.48 kg/ha, indaziflam at 0.033 kg/ha, and pendimethalin at 3.35 kg/ha. At 3 weeks after seeding (WAS) when herbicide was not applied, *Coreopsis lanceolata* had 12 plants dm^{-2} which was over twice as many emerged plants as *Rudbeckia fulgida* and *Chamaecrista fasciculata* and four times as many emerged plants as *Monarda fistulosa*. Pendimethalin did not reduce plant emergence of any of these species, indaziflam did not reduce emergence of *R. fulgida* or *C. fasciculata*. Diuron reduced emergence of all species. At 10 WAS, pendimethalin reduced height of *C. fasciculata* from 15 to 8 cm but did not affect height of other species. Indaziflam reduced height of *C. fasciculata* and *M. fistulosa*. Diuron reduced height of all species. When no charcoal was used, plant emergence 3 WAS was generally reduced compared to charcoal-treated pots. When 672 kg/ha charcoal was applied half OT and half in furrow, plant emergence was numerically highest and statistically equivalent to nontreated pots in all cases. These data suggest that a small increase in establishment rate can be achieved by using 672 kg/ha but 336 kg/ha charcoal would also be a viable treatment. Pendimethalin or indaziflam appear to be effective herbicides for pollinator plant establishment but charcoal banding would be recommended for best results.

UPDATE ON 2018 WEED SCIENCE RESEARCH IN THE IR-4 ENVIRONMENTAL HORTICULTURE PROGRAM. C. Palmer* and E. Vea, IR-4 Project, Princeton, NJ (86)

ABSTRACT

In 2018, the IR-4 Ornamental Horticulture Research Program sponsored research on several weed science projects: pre- and postemergent herbicide crop safety with over-the-top in-season applications, efficacy of post-emergent herbicides, and Christmas tree crop safety with herbicides used to manage cover crops. For the over-the-top herbicide crop safety screening the goal was to develop lists of crops where applications would not harm woody and herbaceous perennials grown in container nurseries. Products tested included Basagran T/O 4F (bentazon), Biathlon (oxyfluorfen + prodiamine), Dimension 2EW (dithiopyr), Dismiss 4F (sulfentrazone), F6875 4SC (sulfentrazone + prodiamine), Fiesta herbicide (FeHDTA), Fortress (aka OHP1701B, isoxaben + dithiopyr), Freehand G (pendimethalin + dimethenamid-p), Gallery SC (isoxaben), Gemini Granular (prodiamine + isoxaben), Pendulum G (pendimethalin), SP1770, and Tower EC (dimethenamid-p). Applications were made at dormancy (preemergents) or when leaves were fully expanded (postemergents) and then approximately 6 weeks later for all products. Basagran was screened on 14 crops; Biathlon on 27; Dimension EW on 4; Dismiss on 11; F6875 4SC on 1; Fiesta on 16; Fortress on 5, Freehand on 16; Gallery on 4; Gemini Granular on 5; Pendulum G on 4; SP1770 on 3; and Tower was screened on 3 crops. The results from this research will aid in the development of product labels and will help growers and landscape care professionals make more informed product choices.

LIVING ACRES: A BIODEVERSITY RESEARCH INITIATIVE THAT IS HELPING LANDOWNERS CREATE A SPACE FOR MONARCH BUTTERFLIES. K.E. Kalmowitz* and H. Coble, BASF Corp, Raleigh, NC (87)

ABSTRACT

Monarch butterfly populations have been declining in the United States since the late 1990's. One of the many factors contributing to this decline is the shrinking number of milkweed plants. Milkweed is the critical component in the monarch's reproduction cycle.

Most monarchs fly north from central Mexico in spring and summer, landing in central and eastern parts of the United States. During this migration, adult female monarchs lay eggs only on milkweed plants. When the eggs hatch, the resulting larvae feed on milkweed leaves. Thousands of golf courses lie along the monarchs' migration path. By allowing milkweed species to grow on land in non-play areas on golf courses, superintendents can play a key role in helping maintain and restore the monarch population. However, common milkweed seedlings are susceptible to many commonly used selective and non-selective herbicides as well as mowing and mechanical disturbance, common maintenance in out-of-play golf courses areas. Superintendents in recent years have been evaluating different practices that will reduce labor and chemical and fertilizer inputs in these non-play areas. Unlike so many other environmental challenges, the milkweed population decline has a straight-forward solution. It's as simple as establishing a butterfly habitat that includes milkweed and nectar plants. The Living Acres Project provides guidance on how superintendents can establish habitat to help preserve this unique and beloved butterfly species and at the same time accomplish savings in overall maintenance. Participants receive directions on what sites around the golf course provide best habitat, suggested milkweed species for specific geographies and how to plant and perpetuate their milkweed stands for years of use for migrating butterfly. The Living Acres Project has demonstrated land throughout the golf course, in non-play areas, can support milkweed communities without interfering with the sport or operations.

INVASIVE VASCULAR PLANT SPECIES AT J.N. DING DARLING NATIONAL WILDLIFE REFUGE, SANIBEL ISLAND, FLORIDA. R. Stalter* and E. Lamont, St Johns University, Queens, NY (91)

ABSTRACT

The objective of this study was to collect and document the invasive vascular plant species at 2104 hectare J. N. Ding Darling National Wildlife Refuge, Sanibel Island, Florida, 26 27' 30"N, 82 06' 00"W. Vascular plant species were collected at 2 month intervals from May 2014 to October 2017 during which we identified 328 species in 250 genera in 92 families Seventy five species, 23% of the flora were not native to the region .Most invasives were found in the ruderal community, one of six arbitrarily classified communities at the refuge. Ten invasive vascular plant species were found in the *Fabaceae*, the second largest family in the flora (31 taxa). *Urochloa* (*Poaceae*) was the genus with the greatest number of invasive taxa, 3, followed by *Dioscorea* and *Agave* each with 2. U.S. Fish and Wildlife Service personnel have removed the most aggressive invasive, *Schinus terrebinthefolius*, from the refuge, documentation that invasive taxa may be controlled within preserves, refuges and parks. One rare taxon, *Celtis pallida*, is present at the shell midden hammock community.

MDOT SHA LANDSCAPE MANAGEMENT GUIDE - A PERFORMANCE BASED GUIDE FOR MANAGING GREEN ASSETS ALONG MARYLAND HIGHWAYS. J.M. KROUSE*, MARYLAND DEPARTMENT OF TRANSPORTATION STATE HIGHWAY ADMINISTRATION, BALTIMORE, MD. J.K. Krouse*, MDOT - SHA, Baltimore, MD (92)

ABSTRACT

The Maryland Department of Transportation State Highway Administration (MDOT SHA) published *Integrated Vegetation Management Manual for Maryland Highways* (IVMM) in 2003. The IVMM provided comprehensive guidance for roadside management, and was the primary policy and training guide for mowing, weed control, and other concerns related to turf, brush control and recordkeeping. The IVMM provided standards used by State employees and contractors, and was often adopted by reference by other government agencies.

Since publication of the IVMM, technical improvements and concerns about the costs and consequences of management operations; interest in native meadow vegetation; declining populations of honeybee (*Apis mellifera* L.), monarch butterfly (*Danaus plexippus* L.) and other pollinator species all provided reasons for MDOT SHA to refocus its management approach to the roadside landscape. Instead of an updated guide to mowing and spraying, the *MDOT SHA Landscape Management Guide* (LMG) was developed as an entirely new document and full replacement of the IVVM.

The LMG prioritizes the delineation of distinct corridor management zones (CMZ's) with unique zone goals. Rather than providing detailed operational guidance, the LMG presents management objectives as ways to meet performance based goals of the CMZ's. In this way, the LMG emphasizes highway safety and operability while providing a streamlined set of management operations. Throughout, the intent of the LMG is to promote a green asset oriented view of management that focuses on meeting zone goals, while deliberately refocusing the management mindset towards a more 'nurturing' view of highway vegetation.

Key concepts of the LMG were delivered to highway managers in a series of presentations during 2018, and the LMG was released as a test draft pending final adjustments and approval. The emphasis on 'nurturing' appeared to resonate strongly with diverse audiences, and suggests that vegetation management guidance that emphasizes positive outcomes may significantly enhance the acceptance of improved practices.

COOPERATIVE WEED CONTROL PROGRAMS AND THEIR IMPACT ON NOXIOUS AND INVASIVE WEEDS IN MARYLAND. L.K. Heimer*, Maryland Department of Agriculture, Annapolis, MD (93)

ABSTRACT

The Maryland Department of Agriculture is responsible for the enforcement of Maryland's Noxious Weed Law. The Noxious Weed Law has a provision that the Maryland Department of Agriculture may enter into a cooperative agreement with a county or political subdivision to provide management, technical assistance, training, and education for implementing a noxious weed control program. These county programs are now the core of the noxious weed control efforts in Maryland.

As a result of the possible addition of Palmer Amaranth to the list, an evaluation of the county cooperative programs was initiated to determine the success of the county programs. Current and past activities were assessed to make recommendations for focusing and improvement. Assessment of noxious weed control in Maryland as it relates to county program size and management.

WHAT IS THE CURRENT KNOWN STATUS OF WAVYLEAF BASKETGRASS (*OPLISMENUS UNDULATIFOLIUS*) IN PENNSYLVANIA? A. Gover*, E.B. Euker, and E.B. DePaulis, Penn State University, University Park, PA (94)

ABSTRACT

Wavyleaf basketgrass (*Oplismenus undulatifolius* (Ard.) P. Beauv.; OPLUN) is a shade tolerant, cool-season, stoloniferous, perennial grass first confirmed in the United States in 1996 in Patapsco Valley State Park, Maryland, and in southeast Pennsylvania in Codorus (CODO) and Gifford Pinchot (GIPI) State Parks, both in York County, in September 2016. Surveys in other SE PA State Parks in 2017 and 2018 have been negative. Additional sites within GIPI were discovered in 2017, and one additional site at CODO in 2018. To date, there has only been one other OPLUN sighting in PA, also in York County at Apollo County Park, which was reported on iNaturalist.org. One challenge is keeping abreast of at least three citizen science reporting sites, iNaturalist, EDDMapS, or iMapInvasives. We conducted a workshop June 2018 at GIPI where participants engaged in survey and herbicide treatment. A key point for participants was the apparent random distribution of OPLUN throughout the work area, as it frequently occurred off-trail in areas seemingly untrafficked by people or wildlife. With no discernible pattern, surveyors will need to engage in fine-scale survey if they wish to find infestations early. Through 2017 Pennsylvania sightings were treated with a glyphosate-based herbicide mixture as they were discovered. Glyphosate-treated sites were populated with seedlings in 2018, leading us to question if using a graminicide would leave more vegetation to provide competition, particularly since many of the treated locations were populated with native understory species. We used quizalofop-P ethyl at 73 g/ha through the summer, but were not satisfied with the efficacy, and reverted to glyphosate for some of the late-season applications. We are currently planning on collaborating beginning 2019 with Towson University to conduct multi-season control experiments at sites in MD. A key objective will be to determine which graminicides have the most activity, as well as test approaches to reduce reinfestation from seed such as combining preemergence herbicides with foliar treatments. OPLUN is regulated in PA, listed as a "Class A" Noxious Weed, accompanied by the language "Preventing new infestations and eradicating existing infestations of noxious weeds in this class is high priority." We will continue to post updates on an OPLUN-specific listserve created in 2017, and provide updates at Cooperative Extension venues during the winter of 2019. We are working to optimize effort, as we are still not confident in our assessment of the threat level of OPLUN compared to other exotic species. If OPLUN is an exceptional invader, we want to prevent widespread infestation. If it's just another invader, we do not want to reduce other useful efforts.

TIMING PRE-EMERGENT CONTROL OF JAPANESE STILTGRASS: FIELD TRIALS FOR EFFICACY OF INDAZIFLAM AND PENDIMETHALIN. B.F. McDonnell*, National Park Service, Bushkill, PA (95)

ABSTRACT

Two field trials were established to test the efficacy of indaziflam and pendimethalin as pre-emergent treatments for *Microstegium vimineum*, Japanese stiltgrass. The first, from spring, 2017, was established approximately 3 to 4 weeks before expected stiltgrass germination. The single trial plots for each component were large, approximately 0.2 acres. Pendimethalin showed very good control at this timing, indaziflam control was poor. Suspecting trial start time may be critical for effective control, a small plot trial with 4 replicates of each trial component was established in October, 2017, and efficacy was assessed during the 2018 growing season. Through the early the growing season, May through July, both pendimethalin and indaziflam resulted in good control. However, by August, indaziflam continued to exhibit in good control, but pendimethalin control declined. Conclusion: both pendimethalin and indaziflam can provide good pre-emergent control of *Microstegium vimineum*, but timing of application is important. Pendimethalin can be applied closer to germination, but indaziflam will provide a longer period of control, if it is applied early enough to be incorporated.

IS INDAZIFLAM A VIABLE ALTERNATIVE TO EXISTING HERBICIDES FOR JAPANESE STILTGRASS (*MICROSTEGIUM VIMINEUM*) MANAGMENT? A. Gover*, A.C. Causey, and M.L. Fuchs, Penn State University, University Park, PA (96)

ABSTRACT

Japanese stiltgrass (*Microstegium vimineum* (Trin.) A. Camus) is a summer annual, C_4 grass native to east Asia, widely considered problematic in wildland settings. Indaziflam at 37, 51, 73, or 102 g/ha has been tested on stiltgrass as a spring or fall PRE treatment, as well as EPOST in tank mixes against herbicides with established efficacy such as pendimethalin or prodiamine. The following experiments were randomized complete block designs with four replications, and percent total vegetation and stiltgrass cover were visually assessed. Data were subjected to analysis of variance, without transformation. In 2016, Indaziflam applied March 23 (37, 51, 73, 102 g/ha) PRE or June 1 EPOST (51 or 73 g/ha) to a sparsely vegetated site at Reeds Gap State Park, Milroy, PA, had stiltgrass cover ratings ranging from 13 percent (37 and 51 g/ha PRE), 5 percent (51 and 73 g/ha EPOST), to 2 percent (73 and 102 g/ha PRE) on August 4, while the untreated plots had 27 percent stiltgrass cover (33 percent total cover) and prodiamine at 1.46 kg/ha (PRE) was rated at 1 percent cover. In 2017, treatments were compared on a more productive site on Penn State Forestland near Petersburg, PA. Indaziflam was applied PRE at 37, 51, 73, or 102 g/ha on March 9, and the same rates EPOST on May 9 in combination with fenoxaprop-P at 50 g/ha. Pendimethalin was applied PRE at 2.3 kg/ha, and prodiamine was applied EPOST at 1.64 kg/ha in combination with flumioxazin at 36 g/ha or imazapic at 34 g/ha. On August 23, untreated plots averaged 100 percent total vegetative cover, and 58 percent stiltgrass cover. Indaziflam PRE at 37, 51, 73, or 102 g/ha averaged 30, 16, 44, and 17 percent stiltgrass cover, respectively, while pendimethalin was rated at 9 percent (100 percent total cover). The indaziflam EPOST combination treatments at 37, 51, 73, or 102 g/ha were rated at 10, 5, 2, and 1 percent stiltgrass cover, respectively, and fenoxaprop-P alone at 50 g/ha was rated at 11 percent stiltgrass cover. EPOST prodiamine treatments were rated at 1 and 7 percent stiltgrass cover for the flumioxazin or imazapic combinations, respectively. The 2018 evaluations were conducted on the edge of a gravel road in the Penn State Forestlands, near Petersburg, PA. PRE treatments of indaziflam at 51, 73, or 102 g/ha or pendimethalin at 2.3 kg/ha were applied in the fall (October 23, 2017) or spring (March 6, 2018). Pendimethalin at 4.4 kg/ha or prodiamine at 1.6 kg was also applied at the fall timing. On July 12, 2018, the untreated plots averaged 75 percent total cover and 59 percent stiltgrass cover. Indaziflam treatments provided similar results whether applied fall or spring, with the 51, 73, and 102 g/ha treatments applied fall were rated at 18, 23, and 3 percent stiltgrass cover, respectively, and the spring treatments were rated at 31, 25, and 10 percent, respectively. Pendimethalin at 2.2 kg/ha was rated at 1 and 4 percent stiltgrass cover, for the fall and spring applications, respectively. Fall applications of pendimethalin at 4.4 or prodiamine at 1.6 kg/ha were rated at 0 and 2 percent stiltgrass cover, respectively. Indaziflam provides significant suppression of stiltgrass at the higher tested rates, and has consistently been statistically indistinguishable from prodiamine and pendimethalin.

NEWSS YEAR-END FINANCIAL STATEMENT 2018

AUDIT REPORT – Kurt Vollmer

Financial documents and bank accounts were audited by,
Quintin Johnson and Barbara Scott on December 12, 2018

BEGINNING STATEMENT: As of October 31, 2017, the NEWSS checking account showed a balance of **$65,956.35**, while the savings account balance was **$31,850.92**. The endowment fund balance was **$26,194.01**

ENDING STATEMENT: As of October 31, 2018, the NEWSS checking account showed a balance of **$61,783.14**, while the savings account balance was **$31,857.27**. The endowment fund balance was **$29,786.33**

Net gain in assets for the fiscal year: $123,426.74-$124,001.28 = -$574.54

FISCAL YEAR ACCOUNTING

New income as shown on statements for November 2017 through October 2018 including earned interest and endowment gain was **$53,829.65**

New deductions from checking for November 2017 through October 2018 totaled $**54,404.19**

Net gain *in assets:* ***$53,829.65 - $54,404.19 = -$574.54***

Non Annual Meeting Revenue/Expenses:

Non annual meeting revenue, which was comprised of interest earned as well as member dues only, sustaining members (10,000), and other contributors totaled **$16,758.67**

Non annual meeting expenses (Dishonesty bond, website, administrative costs, Dir. Science Policy (2018), endowment fund investment, hotel fines, and CAST dues for 2018 and 2019 totaled **$14,815.60**

Annual Meeting Revenue/Expenses:

Annual Meeting revenue (registration, industry social, proceedings residuals, quiz bowl donation, 2018 Hotel Refund) totaled **$37,070.98**

Annual meeting expenses (hotel + AV, PayPal fees, profit distribution for 2017 and 2018, programs, awards, meeting management, honorariums, and 2019 hotel deposit) totaled **$39,588.59**

Education Fund Contributions

Contributions received for the Education Fund include proceeds from sales of *Weeds of the Northeast* ($525) totaled **$1,760** for the 2018 fiscal year.

NEWSS FISCAL STATEMENT
NOVEMBER 1, 2017- OCTOBER 31, 2018

Income Received	Amount
Joint Meeting Total Registration	$27,020.00
Member Dues Only	$250.00
Industry Social (3 Sponsors at $3000.00)	$9,000.00
Interest earned on savings account	$6.35
Net gain in endowment	$3,592.32
Sustaining Membership: 12 to date	$10,000.00
Additional Support for 2017 Weed Contest	$1,000.00
Endowment Contributions	$1,760.00
Weed Contest Hotel Cleanup	$150.00
Proceedings Residuals	$96.00
NEPPSC Quiz Bowl Donation	$210.00
2018 Hotel Refund	$744.98
Total NEWSS Income Nov 1, 2017- October 31, 2018	**$53,829.65**

Expenses Paid	Amount
PayPal Fees	$877.90
Joint Meeting Hotel + AV Cost	$24,467.28
Joint Meeting Profit Distribution 2017	$3,785.99
Joint Meeting Profit Distribution 2018	$1,469.68
Sustaining Member Poster	$242.81
Alphagraphics - Mtg Programs	$479.96
NEWSS Award Booklets	$153.43
2018 Speaker Honoraria	$958.29
EC Dishonesty Bond Insurance Renewal	$359.00
Website "A Small Orange" domain fee for NEWSS org.	$228.00
New NEWSS/NEPPSC website	$4,500.00
Administrative	$196.10
Meeting Management	$2,255.25
Quiz Bowl-Amazon Gift Cards	$210.00

Annual meeting Membership Awards - Etch Art	$413.00
Annual Meeting Poster Awards (1st and 2nd)	$300.00
Annual Meeting Student Paper Awards (1st and 2nd place)	$300.00
Annual Meeting Photo Contest (1st, 2nd and 3rd place)	$175.00
2017 Endowment Investment	$3,682.50
Weed Contest Hotel Fines	$250.00
WSSA Dir. Science Policy YEAR 2018 Dues	$2,600.00
CAST 2018 dues	$1,500.00
CAST 2019 dues	$1,500.00
2019 Hotel Deposit	$3,500.00
Total NEWSS Expenses paid Nov 1, 2017- October 31, 2018	**$54,404.19**

Total Fiscal Year Summary	
Total Income	$53,829.65
Total Expenses	$54,404.19
Net Gain	**-$574.54**

NEWSS Account Balances	*October 31, 2018*
Checking	$61,783.14
Savings	$31,857.27
Education Fund	$29,786.33
Total Net Assets	**$123,426.74**

NORTHEASTERN
WEED SCIENCE SOCIETY

2018 NEWSS BUSINESS MEETING

Call to Order

President Randy Prostak called the meeting to order at 1637.

President's Comments

President Prostak provided a brief recap of the meeting attendance.

Offered kudos to Kate Venner, Kurt Vollmer, Sudeep Matthew, and Quintin Johnson for their efforts upgrading the Society website.

Reminded the Society to share input on the ongoing status of NEPPSC.

Student Endowment Fund is close to $33,000. Interest revenue was tapped for 2018 Quiz Bowl awards.

Praised the Society's can-do attitude, and helpfulness, in general, and specifically during his tenure in the Presidential arc.

Encouraged members to serve on the Board.

Acceptance of the 2017 Business Meeting Notes

A few minutes were taken to review the minutes from the 2017 Business Meeting. Jeff Derr moved to approve, Dave Yarborough seconded. Minutes approved without revision.

Membership Chair Report - Sudeep Matthew

MEMBERSHIP

Membership report distributed, with 10-year summary. After a peak of 140 Regular Members in 2016, numbers decreased to 96 then 65 in 2017 and 2018 respectively. Sudeep reminded us that it is up to us as members to encourage other members, and that small numbers matter, as they add up in a Society of our size.

NECROLOGY

Henry P. Wilson, 1/7/17. Society President 1975-76.

Treasurer's Report - Kurt Vollmer

Kurt reviewed submitted report.

Audit was conducted 12/11/17 by Todd Davis and Quintin Johnson. The accounts were found to be in order.

Total Fiscal Year Summary	
Total Income	$62,915.17
Total Expenses	$47,875.71
Net Gain	**$15,039.46**

NEWSS Account Balances, 10/31/2017	
Checking	$65,956.35
Savings	$31,850.92
Education Fund	$26,194.01
Total Net Assets	**$124,001.28**

Education Fund was at $26,194.01 as of October 31, but closer to $35,000 now.

- the first use of revenue, from interest, was Amazon gift cards for 2018 Quiz Bowl winners.
- cash prizes are being considered for 2018 Weed Contest winners.

BASF is hosting 2018 Weed Contest.

Encouraged Society input on usage of NEPPSC meeting funds.

Awards Committee - Shawn Askew

RECAP OF 2018 SOCIETY AWARDS

GRADUATE STUDENT PRESENTATION AWARDS – Rakesh Chandran

Review of comments - overall very good this year; "Tell your story", limit details to the essentials, don't lose your audience in the details, clearly state objective(s) and hypothesis; readable font size; uncluttered charts and graphics; limit scope of conclusions to the data provided; be clear about statistics and experimental design; preliminary explanation of graph/chart set-up; and use SI units.

Shawn adds - only share information relevant to the particular experiments; introduction should be complete within 5-6 minutes.

Presentation of awards

GRADUATE STUDENT POSTER AWARDS - Matt Elmore

Review of comments - overall quality was good this year; follow the rubric on the website; poster should be stand-alone; make sure figures are needed; proofread, proofread, proofread.

Presentation of awards

Photo Contest - Art Gover

There were 22 total entries from 6 contestants. Art reminded the Society that the Photo Contest is an easy way for members to be in the running for real money, doing something they do as a matter of course.

Third Place - Sandeep Rana - panoramic view of a mustard field

Second Place - Eugene Law - blooming common teasel

First Place - Thierry Besançon - bloom of Carolina redroot

Weed Science Policy - Lee Van Wychen

Lee provided a quick update (yes, really) on the following topics:

- WSSA EPA Liaison - Greg Kruger has assumed this role, taking over for Mike Barrett.
- USDA-NIFA Fellow - Don Schilling is stepping down after three years of service. WSSA will need to appoint his successor.
- USDA Leadership Vacancies - Lee reported there is an effort to get Rich Bonanno appointed to the role of Under Secretary for Research, Education, and Economics; a position regarded as USDA's chief scientist.
- FY 2018 Federal Budget - we are currently working under a Continuing Resolution through January 19, with many issues unresolved in working towards an actual budget. Ag appropriations are at satisfactory levels, provided sequestration caps are not implemented.
- Dicamba - EPA announced label changes to the dicamba products Extendimax, Engenia, and Fexapen on October 13, 2017. EPA has indicated that if the issues encountered in 2017 are repeated, 2019 registration is questionable.
- Glyphosate - WSSA in on record stating "the IARC review process for glyphosate was flawed and represents a case of gross scientific negligence. There is no question that IARC arrived at their conclusion due to their inclusion of the positive findings from a selection of studies with known limitations, a lack of reproducible positive findings, and the omission of the negative findings from credible and reliable research." Lee also summarized the November 9, 2017 results in the Agricultural Health Study published by the National Cancer Institute; and EPA's December 19, 2017, human health draft risk assessment.
- EPA Herbicide Resistance Guidance - Referred to as Pesticide Registration Notice 2017-2, this document will apply to all herbicide uses except residential settings (lawns).
- National Invasive Species Awareness Week (NISAW) - February 26 - March 2, 2018.
- Farm Bill - Congress will begin working on the updated Farm Bill in 2018.
- 2018 Weed Survey - focus will be on weeds in non-crop areas.

CAST Representative Report - Carroll Moseley (for Rob Richardson)

Summarized 2017 publications, and topics in the proposal and development stage, and new leadership of the Plant Working Group (compiled by Rob Richardson).

Officer Changeover and Presentation of the NEWSS Gavel

Randy Prostak passed the NEWSS presidency to Carroll Moseley, and presented The NEWSS Gavel.

Carroll's first order of business was to present the ceremonial plaque and gavel to now Past-president Randy Prostak.

New Business - Carroll Moseley

NOMINATION COMMITTEE

The Committee offered Art Gover of Penn State as candidate for Vice-president. President Moseley opened the floor for additional nominations. Jeff Derr moved and Dave Yarborough seconded that nominations be closed. Art Gover approved as new Vice President.

Carroll Moseley appointed Randy Prostak and Jeff Derr to the 2018 committee. Randy Prostak nominated Sudeep Matthew. Michael Flessner nominated Shawn Askew. Anthony Witcher nominated Thierry Besançon. Approved by acclamation.

Jerry Baron, Matthew Elmore, and Matt Cutulle were nominated to Resolutions Committee.

Membership Chair - Stacie Pickard nominated, approved by acclamation.

WSSA Representative - Rakesh Chandran was appointed.

OTHER BUSINESS

Society Editor Michael Flessner reminds us that the abstract submission feature on the wssaabstracts website is now open for delinquent abstracts.

Next year's meeting dates and venues are not yet final, but will be announced within 30 days.

Collegiate Weed Contest is at Pine Level, NC, hosted by BASF, July 23-25, 2018.

Carroll reminded us that NEWSS is an exemplary setting for graduate student development and camaraderie, and comparatively unique compared to other societies.

Presentation of the NEWSS Board

The board was presented, with group photograph to follow adjournment.

Adjourn

Shawn Askew moved, Sudeep Matthew seconded, motion approved at 1801.

Author Index